Solid Modelling with AutoCAD

Solid Modelling with AutoCAD

Robert McFarlane MSc, BSc, ARCST, C.Eng, MIMech E, MIEE, MILog

Senior Lecturer, Department of Integrated Engineering, Motherwell College

Edward Arnold
A member of the Hodder Headline Group
LONDON   SYDNEY   AUCKLAND

To Peter, Ian, Derrick, Alex, Craig, Ronnie, Chris, Callum, Craig, Robert and Alex – my first full-time HNC CAD class

First published in Great Britain 1995 by
Edward Arnold, a division of Hodder Headline PLC,
338 Euston Road, London NW1 3BH

Whilst the advice and information in this book is believed to be true and
accurate at the date of going to press, neither the author nor the publisher can
accept any legal responsibility or liability for any errors or omissions that may
be made.

British Library Cataloguing in Publication Data
A catalogue record for this book is available from the British Library

ISBN 0 340 63204 6

1 2 3 4 5   95 96 97 98 99

Produced by Gray Publishing, Tunbridge Wells
Printed and bound in Great Britain by The Bath Press, Avon

Contents

Preface   vii

1. Model types   1
   Wire-frame modelling   Surface modelling   Solid modelling

2. The solid model standard sheet   4
   Checking your standard sheet   Revision

3. Exercise 1: positioning block   12
   Summary of commands

4. Exercise 2: positioning block – profile and
   dimensioning   18
   Summary

5. Exercise 3: the BASIC primitives   21
   *The BOX primitive   The WEDGE primitive
   The CYLINDER primitive   The CONE primitive
   The SPHERE primitive   The TORUS primitive
   Baseplanes and construction planes*

6. Exercise 4: the SWEPT primitives   32
   The extrusions   The revolutions   Summary of commands

7. Exercise 5: the EDGE primitives   43
   The CHAMFERED edges   The FILLETED edges   Summary of commands

8. Boolean operations   53
   *The union operation   The difference operation
   The intersection operation   Separation of solids   Cutting solids
   The CSG tree   Balanced and unbalanced trees
   Activity – predicting Boolean operations*

9. Exercise 6: a machine support   60

10. Exercise 7: a backing plate   69
    The model sizes   Investigating the composite

11. Exercise 8: a sort of level   76
    Traditional 2D drawing   Solid model

12. Exercise 9: pipe and flange   80

13. Review   83
    *Classic geometry primitives   Swept primitives   Edge primitives
    Irregular surfaces   Thin shells   Efficient modelling*

14. Regions   84
    Region examples

15. Moving solids   92
    *The motion co-ordinate system (MCS)   .SOLMOVE example 1
    Motion description codes   Response descriptions   SOLMOVE example 2
    Summary*

16. Changing solids   99
SOLCHP example 1   SOLCHP example 2

17. Exercise 10: a casting block   105
The basic casting   Modifying the casting block

18. Solid modelling variables   112
*Display   Model operations   Units   Materials and mass property analysis
2D view extraction   Miscellaneous   Checking the solid model variables
The SOLVAR dialogue box   Summary*

19. Material properties   116
*SOLLIST   SOLAREA   SOLMASSP   SOLMAT   Write to file
Summary*

20. Exercise 11   122
Analysis of composite   Entering our own Material Values

21. Exercise 12: extracting sections   131

22. Exercise 13: a desk tidy as a detail drawing   134
*Traditional 2D drawing   The basic composite
The compartments   The front cut out   The holes
The auxiliary feature extraction   The section extraction
Extracting the profiles   Tidying up the drawing   Moving the viewports
Dimensioning the drawing   Summary*

23. Exercise 14: a computer link detail drawing   146
*Starting the drawing   Primitive 1   Primitive 2   Primitive 3
Primitive 4   Primitive 5   Final composite   New viewports
Extracting two new views   New layers
Extracting the features for the auxiliary and the true shape
Erasing unwanted detail   Viewport alignment
Dimensioning the drawing*

24. Exercise 15: block assembly   157
*The tray   The support   Insert 1   Insert 2   'Cutting' the assembly
Sectioning the assembly*

25. User exercises   168
*Pipe bend 1   Tail pipe   Double flanged pipe bend   Pipe bend 2
Pipe bend 3   Greek temple   A casting   Cone/cylinder interpenetration
Elliptical pipe   A thin-walled shell   SOLBOX structure
Newton's cradle*

26. Shading solid models   182
Shading   Summary

27. DVIEW with solid models   189
*Getting started   Adding detail   Dynamic viewing the house
The DVIEW options   Creating a slide show   Summary
An extra solid model composite for DVIEW   DVIEWing the composite*

28. Rendering solid models   201
*Loading AutoCAD render   AutoCAD's render 'pictures'
Render example   Lights   Scenes   Images   Finishes*

29. . . . and finally   209

Index   211

Preface

I started using AutoCAD in 1985 with Version 2.something. Since then, developments in the software combined with technological advances in the hardware have resulted in Releases which have made AutoCAD into a very powerful PC-based draughting and design package. Users now have access to topics which a few years ago were only possible on mainframe/workstations, such as

- 3D wire-frame and surface modelling
- customisation
- rendering
- solid modelling.

Solid modelling with AutoCAD is relatively simple to learn, and users will be quite amazed at what will be produced with only a few hours of tuition. This book is intended for *all AutoCAD users* who want to learn solid modelling and will benefit both the specialist and the beginner as well as undergraduates and post-graduates who require solid modelling in their courses. This is an interactive book and the user can complete exercises from information supplied. It is my intention to show the versatility of solid modelling with as many diverse examples as possible. I have also introduced sections on shading and rendering which the user will find very interesting and useful. The book covers the work of the SCOTVEC Higher National Unit in Solid Modelling which is a core unit in the HNC Computer-aided Draughting and Design.

Requirements

The following are considered essential for using the book:

1. AutoCAD Release 12.
2. A 486 PC conforming with the AutoCAD requirements. A 386 machine is satisfactory but it is slow. I completed all the exercises on a 386 machine.
3. Enough memory on the hard drive to store drawings. While a floppy may be used, you may run out of storage space rather quickly.
4. Readers should have:

   (a) a good working knowledge of AutoCAD, i.e. commands, dialogue boxes
   (b) a sound grasp of 2D and 3D draughting concepts, especially UCS, Viewports, Paper Space and Model Space.
   (c) the ability to want to learn.

Using the book

As the book is interactive there are several concepts with which the reader must be conversant with:

1. User input will be in **bold** type, e.g.
   - enter **SOLBOX** <R> from the keyboard
   - pick **Model** from the screen menu
   - pick **File** from the menu bar.
2. Prompts from AutoCAD will be in `typewriter` type.
3. AutoCAD prompts will be give in full when commands are being used for the first time.
4. The ENTER or RETURN key will be denoted by **<R>** or **<RETURN>**.
5. The user is recommended to make a directory to save all drawings, and it is *strongly advisable* to save drawings regularly.

1. Model types

The word 'model' can convey different concepts to different people and the following give some examples of this difference:

- *Physical models:* which may include sculptures, prototypes, mock ups.
- *Flat models:* which could be photographs, sketches.
- *Computer models:* two- and three-dimensional (2D and 3D) image displays.
- *CAD models:* with CAD, a model is a mathematical representation of a geometric shape which is stored in the computer memory as a sequence of numbers. There are two basic type of CAD model:

    2D – is a 'flat world' in which the model is usually defined by a sequence of X and Y co-ordinates

    3D – is a 'real world', the model usually being defined by a sequence of X, Y and Z co-ordinates.

3D modelling with CAD can be sub-divided into three categories:

- wire-frame modelling
- surface modelling
- solid modelling.

Wire-frame modelling

Wire-frame models are defined by points and lines and are the simplest possible representation of a 3D component. They may be adequate for certain 3D model representation and require less memory to complete than the other two model types, but wire-frame models have several limitations:

1. Ambiguity: how you view the component, i.e. are you 'looking down at' or 'looking up at' the component.
2. No curved surfaces: while curves can be added to a wire-frame model, an actual curved surface cannot. Lines may be added to give a 'curved effect', but the computer does not recognise these.
3. No interference: as wire-frame models have no surfaces, they cannot detect interference between adjacent components. This makes them unsuitable for kinematic display, simulations, etc.
4. No physical properties: mass, volume, centre of gravity, moments of inertia, etc., cannot be calculated.
5. No shading: it is impossible to 'render' a wire-frame model as there are no surfaces to allow shading, shadows, etc.
6. No hidden removal: again as the model has no surfaces, it is not possible to display it with 'hidden detail' removed.

AutoCAD allows 3D wire-frame models to be constructed.

Surface modelling

A surface model is defined by points, lines and faces and a wire-frame model can be converted into a surface model by adding these 'faces'. Surface models have several advantages over wire-frame models, which include:

1. Recognition and display of curved profiles.
2. Shading, rendering and hidden detail removal possible.
3. Recognition of holes.

Surface models are very suited to many applications but they have some limitations including:

1. No physical properties: other than surface area, a surface model does not allow the calculation of mass, centre of gravity, moments of inertia, radii of gyration, etc.
2. No detail: a surface model does not allow section detail to be obtained.

Several types of surfaces exist which generate a surface model, and these include:

    (a) plane and curved swept surfaces
    (b) swept area surfaces
    (c) rotated or revolved surfaces
    (d) splined curve fitted surfaces – sculptures
    (e) nets or meshes.

AutoCAD allows surface models of all types mentioned.

Solid modelling

A solid model is defined by the volume the component occupies and is thus a real 3D representation of the component. Solid modelling has many advantages which include:

1. Complete physical properties of mass, volume, centre of gravity, moments of inertia, radii of gyration, etc., can all be calculated for a solid model.
2. Full shading, rendering and hidden detail removal.
3. Section views including true shapes can be obtained.
4. Interference between adjacent components can be highlighted.
5. Simulation for kinematics, robotics is possible.

   Solid models are created using a *solid modeller* and there are several types of solid modeller, the two most common being:

1. Constructive solid geometry or constructive representation (CSG/C-REP). The model is created from solid primitives and/or swept surfaces using Boolean operations.
2. Boundary representation (B-REP).

The model is recognised by the edges and faces making up the surfaces, i.e. the *topology* of the component.

   AutoCAD creates solid models using the constructive solid geometry modeller, i.e. the CSG technique.

Comparison of model types

Figure 1.1 shows a simple component (a cuboid with a hole through it) as a wire-frame, surface and solid model – plotted with hidden detail removal.

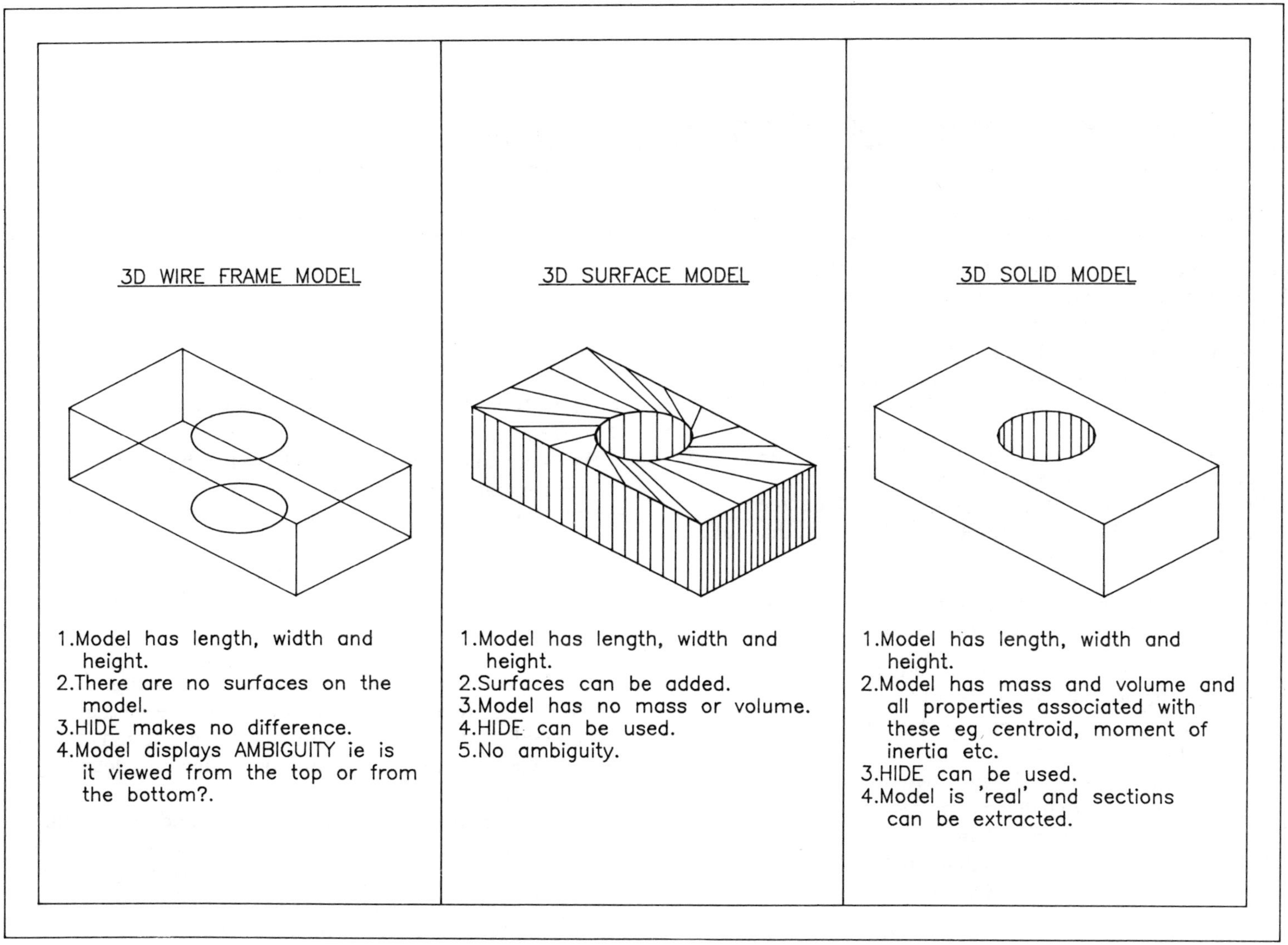

3D WIRE FRAME MODEL

1. Model has length, width and height.
2. There are no surfaces on the model.
3. HIDE makes no difference.
4. Model displays AMBIGUITY ie is it viewed from the top or from the bottom?.

3D SURFACE MODEL

1. Model has length, width and height.
2. Surfaces can be added.
3. Model has no mass or volume.
4. HIDE can be used.
5. No ambiguity.

3D SOLID MODEL

1. Model has length, width and height.
2. Model has mass and volume and all properties associated with these eg centroid, moment of inertia etc.
3. HIDE can be used.
4. Model is 'real' and sections can be extracted.

Fig. 1.1. Simple comparison of wire-frame, surface and solid models.

2. The solid model standard sheet

A standard sheet is required for the exercises in this book. The process of devising the standard sheet is rather long, as there are a number of new concepts which can be introduced to the reader at this early stage, and a revision section has been added at the end of the chapter. As stated earlier on in the book a directory should be used for all your solid model drawings. If you are proficient at this, then go ahead and make one, if you are not sure about this, then follow the sequence below:

1. Switch your system on, and enter the DOS environment – get 'help' if you are unsure of this.
2. Ensure that the **C:>** prompt is showing on the screen.
3. Enter from the keyboard **md\SOLID<R>** – and that's all.
4. Load AutoCAD.

Now that your solid modelling directory has been made, we will proceed with the standard sheet, so:

1. Load AutoCAD if you are not already in it.
2. Set parameters such as BLIPS, APERTURE, GRIPS, etc. to your own requirements, but use the following settings:
   LTSCALE: 0.4
   UNITS: set to decimal with two decimal places, and decimal angles with 0 decimal places.
3. From the menu bar, select **View**

   Tilemode >
   Off(0)

   AutoCAD prompts `New value for TILEMODE<1>:0`
   `Regenerating drawing`

4. You will now be in paper space.
5. Enter **LIMITS<R>**
   AutoCAD prompts `Reset paper space limits...`
   then `Lower left hand corner<0,0>`
   enter **0,0<R>**
   AutoCAD prompts `Upper right corner`
   enter **420,297<R>**
6. Now **ZOOM All** – we have set an A3-sized sheet.
7. *Layers.*
   With our solid models we will have two types of layers:
   (a) general – for the model
   (b) viewport specific – unique to individual viewports.

The general layers will be made with **Settings**
Layer Control...

Layer	Colour	Linetype	Purpose
0	White	Continuous	General use
SOLIDS	Red	Continuous	For the solid models
SHBORDER	Blue	Continuous	Sheet design and text
VPBORDER	Yellow	Continuous	Viewport details

8. Make SHBORDER the current layer and while still in paper space draw the sheet border using the LINE command. The rectangular drawing area is from (0,0) to (380,270). Also draw a line from (0,15) to (380,15) – this will be used be add details of the drawing, e.g. title, date, etc.
9. Now make VPBORDER the current layer.

10. Select from the screen menu **SETTINGS**
    HANDLES

    AutoCAD prompts Handles are enabled (or disabled)
    select **ON**
    AutoCAD prompts Handles are already on (or Handles on)

11. From the menu bar select **View**
    Mview >
    4 Viewports

    AutoCAD prompts Fit/<First point>
    enter **10,25<R>**
    AutoCAD prompts Second point
    enter **370,260<R>**
    AutoCAD prompts: with four equal yellow viewports. These viewports have been drawn within the sheet border.

12. From the screen menu select **INQUIRY**
    LIST

    AutoCAD prompts Select Objects
    respond pick the top right viewport edge and:
    AutoCAD prompts with the text screen giving details about the viewport selected, e.g. layer name, paper size, etc. It also gives the *handle number*. Take a note of this number (my drawing had a handle number of 7 – yours may be different).

13. Repeat the INQUIRY command to obtain the handle number of the other three viewports – my four handle numbers were 7, 8, 9 and A. The A is a hexadecimal number – don't worry about this.

14. Enter model space with **MS<R>** and the 2D icon appears in all four viewports.

15. Set the GRID to 10, and the SNAP to 5 in all viewports.

16. Refer to Fig. 2.1 and using the **VPOINT** command set the following viewpoints:

Viewport	VPOINT	Actual view
top right	0,–1,0	left side
top left	1,0,0	front
bottom left	R 315,30	3D
bottom right	0,0,1	top

Note: some users may prefer to use other methods to 'set' the viewpoint, e.g. the Presets dialogue box, the tripod and axes, etc. My preference is to use the VPOINT command and enter the co-ordinates as above.

17. With the bottom left (3D) viewport active, enter the following sequence to set and save two UCSs:

UCS**<R>**	from keyboard
X**<R>**	rotation about the X axis
90**<R>**	
UCS**<R>**	
S**<R>**	the save option
LEFT**<R>**	name for this UCS
UCS**<R>**	
Y**<R>**	
90**<R>**	
UCS**<R>**	
S**<R>**	
FRONT**<R>**	name for this UCS

Users who are unfamiliar with this sequence, should refer to the Revision section on page 9.

18. These can be checked from the menu bar with
    Settings
    UCS
    Named UCS...

19. Restore WCS.

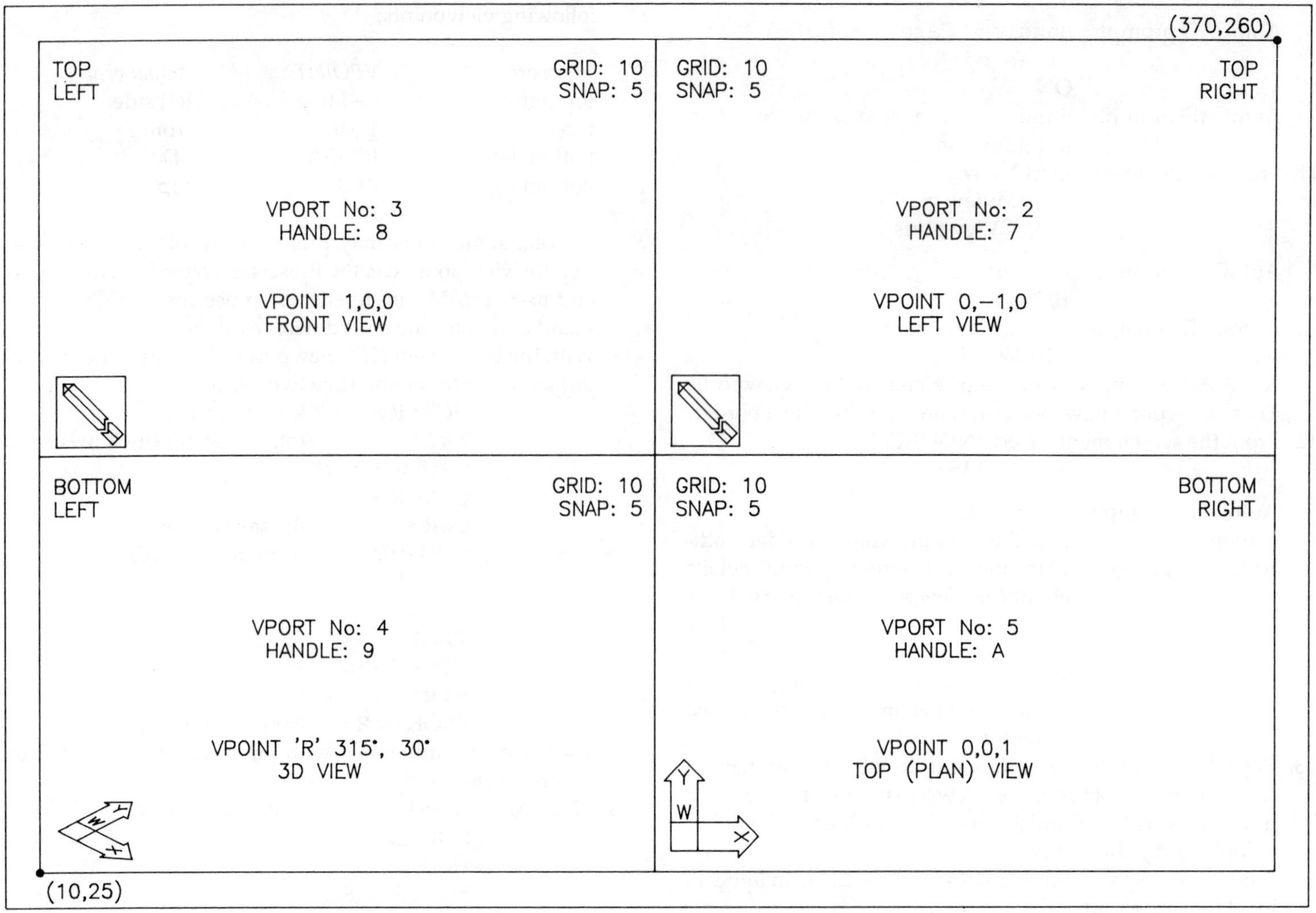

Fig. 2.1. Details of the SOLA3 standard sheet.

20. *Layers specific to the viewports.* We have already made layers for general use, but with solid modelling there are other layers which should be made, to enable us to extract certain details from the model. The specific layers we are interested in are for:
    (a) visible profile detail – PV? layers
    (b) hidden profile detail – PH? layers
    (c) dimension detail – DIM? layers.
       The layers to be created are thus:

    visible profile lines: PV-7, PV-8, PV-9, PV-A
    hidden profile lines: PH-7, PH-8, PH-9, PH-A
    dimensions: DIM-7, DIM-8, DIM-9, DIM-A.

21. Select from the screen menu: **MVIEW**
                                    VPLAYER:

AutoCAD prompts	`?/Freeze/Thaw/Reset/Newfrz/` `Vpvisdflt`
select	**Newfrz**
AutoCAD prompts	`New Viewport frozen layer name(s)`
enter	**PV-7,PH-7,PV-8,PH-8,PV-9,PH-9,PV-A,PH-A<R>**
AutoCAD prompts	`?/Freeze/Thaw.........`
enter	**N<R>**
AutoCAD prompts	`New Viewport frozen layer name(s)`
enter	**DIM-7,DIM-8,DIM-9,DIM-A<R>**
AutoCAD prompts	`?/Freeze/Thaw.........`
enter	**<RETURN>**

Note: 1. Remember that the numbers 7, 8, 9 and A are my handle numbers and that you will have to use your own handle numbers.

2. An explanation of this section is given in the Revision section on page 9.

22. Now select the Layer Control dialogue box, and make:
    (a) all PV layers colour **green**
    (b) all PH layers colour number **9**
    (c) all DIM layers colour **magenta**
    (d) all PH layers linetype **hidden**.

23. The Layer Control dialogue box indicates five possible states for layers:
1. On/Off	On displayed
2. Thaw/Freeze	F displayed
3. Locked/Unlocked	L displayed
4. Current VP Thaw/Freeze	C displayed
5. New VP Thaw/Freeze	N displayed.

    For example:
    (a) OnF... is a frozen layer
    (b) On.... is a thawed layer
    (c) On..C. is a current viewport layer frozen.

24. Enter paper space with **PS<R>**.

25. At the command line, enter **VPLAYER<R>**
AutoCAD prompts	`?/Freeze/Thaw.............`
respond	**pick Thaw**
AutoCAD prompts	`Layer(s) to Thaw`
enter	***7<R>**
AutoCAD prompts	`All/Select/<current>`
respond	**pick Select**
AutoCAD prompts	`Select objects`
respond	**pick the top right viewport<R>**, i.e. handle 7.

26. Repeat the above command and enter *8, *9, *A and pick the corresponding viewport when prompted.

27. Check the above by entering **VPLAYER\<R\>**
    enter   **?\<R\>**
    AutoCAD prompts `Select a Viewport`
    respond   **pick the top right viewport\<R\>**
    AutoCAD prompts `Layers currently frozen in`
    `viewport 2(?)`

    `PV-8  PH-8  DIM-8`
    `PV-9  PH-9  DIM-9`
    `PV-A  PH-A  DIM-A`
    i.e. there should be no layers with number 7

    Cancel the command with **CTRL C**.
28. Now select from the screen menu **DIM:**
                             DimVars
    and set the following dimension variables:

dimaso	OFF	dimgap	2	dimtxt	3	dimtih	OFF
dimasz	3	dimtvp	0	dimscale	0	dimtoh	ON
dimtad	OFF	dimexe	2.5	dimzin	0	dimcen	2

29. From the menu bar select **Draw**
                              Text
                              Set Style
    AutoCAD prompts `Text Font dialogue box`
    respond   **pick Roman Simplex and then OK**
    then   **enter\<R\>** to the six prompts.
30. While still in paper space, make SHBORDER the current layer and add your own sheet requirements using Fig. 2.1 as a guide.
31. Enter model space with **MS\<R\>**.
32. Make SOLIDS the current layer, with the lower left viewport active.

33. Set the solid wire-mesh density by entering **SOLWDENS\<R\>**.
    AutoCAD prompts `Initializing....`
                                 `No modeler is loaded yet.`
                                 `Both AME and Region Modeler`
                                 `are available.`
                                 `Autoload Region/<AME>`
    enter   **\<RETURN\>**
    AutoCAD prompts `Initializing`   `Advanced`
                                 `Modeling Extension`
    then   `Wire-frame mesh density(1 to`
                               `12)<1>`
    enter   **6\<R\>**.
34. And that's your basic solid model standard sheet. Hopefully you have managed to work through these steps without any mistakes. If you have reached this stage, then **SAVE** your standard sheet as:
    \SOLID\SOLA3, i.e. directory name – SOLID
                                      drawing file name – SOLA3
35. This is a standard sheet for all the exercises in this book, and not necessarily a standard sheet which would be used with other solid modelling work. Users may not require the different layers which we have been set.

Checking your standard sheet

1. In model space with the bottom left viewport active, use the LINE command to draw lines as follows:
   (a) from 100,100
       to @100,0
       to @0,0,20
       to @0,50
       to @0,0,–20
       to @0,–50
       to \<RETURN\>

(b) from 100,100
    to @0,0,100
    to @60,0
    to 200,100,20
    to **<RETURN>**
(c) from 100,100,100
    to @0,50
    to @60,0
    to 160,100,100
    to **<RETURN>**
(d) from 200,150,20
    to 160,150,100
    to **<RETURN>**
(e) from 100,100
    to @0,50
    to @100,0
    to **<RETURN>**
(f) from 100,150
    to @0,0,100
    to **<RETURN>**.

2. Make the top right viewport active and use the ZOOM command to 'centre and scale' the view, by entering **ZOOM<R>**

AutoCAD prompts	All/Centre...........
enter	**C<R>**
AutoCAD prompts	Centre point
enter	**150,125,50<R>**
AutoCAD prompts	Magnification or height
enter	**1XP<R>**

3. Repeat the above ZOOM command in the other three viewports, using the same centre point with 1XP in the top left and bottom right viewports, but a value of 0.75XP in the lower left viewport.

4. The component you have just drawn should be aligned correctly as a front, top and left view with a 3D display as shown in Fig. 2.2.

The creation of the solid model standard sheet is involved, but it will save a great deal of time and effort when we are working through the exercises in the book.

Revision

This section will review some of the topics which were used in the creation of the standard sheet. Some users may want to miss this section, although it would do no harm to read through it.

1. Paper space/model space

AutoCAD has multi-view capabilities and this allows the user to layout, organise and plot multiple views of a single drawing. There are two drawing environments available these being:

(a) Model space: is the environment which is used for drawing and is the default setting. Viewports created in model space are called *tiled* and cannot be moved once created. Only the current viewport can be plotted in model space.

(b) Paper space: this environment can be used to add entities to an existing drawing, but the real advantages are:

- viewports are called UNTILED and can be positioned on the paper to the user's requirements
- additional viewports can be added
- any viewport can be erased without interfering with those remaining
- multi-view plotting is obtained with paper space, and selected viewports can have hidden lines removed using Mview-Hideplot.

The two environments can be 'entered' by

(a) selecting View-Paper Space/Model Space from the menu bar or

(b) entering PS or MS at the command line.

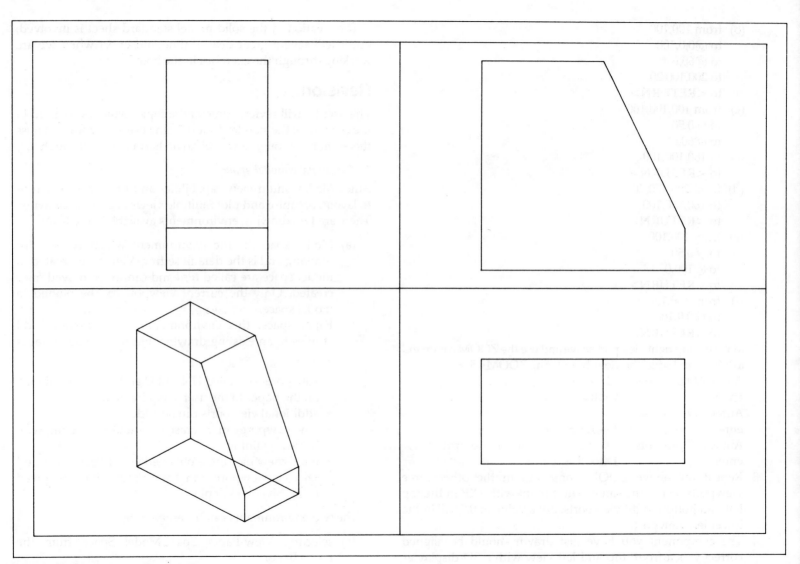

Fig. 2.2. Checking the standard sheet.

2. Tilemode

The tilemode variable controls the drawing environment and:

(a) TILEMODE: 1 – model space with tiled viewports.
(b) TILEMODE: 0 – paper space with untiled viewports.

The variable is activated

(a) from the menu bar with View-Tilemode-Off(0)/On(1) or
(b) by entering TILEMODE at the command line, then 0 or 1.

3. The UCS

The User Co-ordinate System (UCS) allows the user to:

(a) move the origin to any point on the screen
(b) align the icon to any 'plane' on a component
(c) rotate the icon about the X, Y or Z axes
(d) save UCS positions for future recall.

The icon visibility is controlled with UCSICON and can be on/off, set to the origin position in individual or all viewports. In our standard sheet we set and saved two UCS positions using the rotation option about the X and Y axes.

4. VPLAYER

The command VPLAYER (viewport layer) allows the user to control individual viewport layers. Normally when an entity is drawn in one viewport, it is displayed in all other viewports. Similarly when a layer is frozen, all entities on that layer will disappear in every viewport. This is because the layer command is *global* and affects every viewport. VPLAYER only affects the layers in a single viewport, i.e. a named layer can be frozen in one viewport but will be on in the others. This is very useful with multiple viewports and will be used to advantage when we dimension our solid composite models. The options available with VPLAYER are:

(a) *Freeze*: allows named layers to be frozen in one or more viewports.
(b) *Thaw*: thaws frozen layers in named viewports.
(c) *Reset*: restores the default visibility setting of a layer in a named viewport.
(d) *Newfrz*: creates a new layer, frozen in all viewports.
(e) *Vpvisdflt*: defines the layer status of any new viewports which have been created.

Layers created with the VPLAYER command have two status letters which will be displayed in the layer control dialogue box. These are: N which signifies that the layer is a new frozen layer and C which signifies that the layer is currently frozen.

5. ZOOM XP

Views of a component which are created in model space are not normally to a specific scale, and a scaling factor is required to adjust the displayed views in model space before changing to paper space for plotting. The ZOOM XP command allows an accurate scale factor to be set for model space components, and when combined with the centre option, will centre the component in each viewport then scale it full size.

6. Dimvars

Dimension variables (dimvars) control the appearance of dimensions added to drawings. At this level of AutoCAD, most users should be familiar with dimension variables, and can set them to their own specifications as required. The variables set in our standard sheet creation are those which I usually alter in any drawing I create.

3. Exercise 1: positioning block

This first exercise will introduce some of the basic solid modelling drawing commands, as well as those which allow solids to be 'added' and 'subtracted'. All solid modelling commands can be obtained:

1. By entering the command at the prompt line, e.g. **SOLSUB**.
2. By selection from the on-screen menu.
3. From the menu bar.

It is the user's preference as to what method is used, but all three methods will be introduced in this exercise.

1. Load AutoCAD and begin a new drawing **\SOLID\ BLOCK=\SOLID\SOLA3**.
   If your standard sheet was saved successfully, you should be in model space with the lower left viewport active. SOLIDS should be the current layer – red.
2. Refer to Fig. 3.1 which gives the component sizes and details stages in the construction of the solid.
3. From the screen menu, select **MODEL**
   PRISMS.
   SOLBOX

AutoCAD prompts   `Initializing...`
                   `Initializing Advanced Modelling Extension`
then   `Baseplane/Centre/<Corner of box> <0,0,0>`
enter   **0,0,0<R>**

AutoCAD prompts   `Cube/Length/<Other corner>`
enter   **L<R>**
AutoCAD prompts   `Length` and enter **100<R>**
                   `Width` and enter **80<R>**
                   `Height` and enter **40<R>**
AutoCAD prompts   `Phase I - Boundary Evaluation begins`
                   `Phase II - Tessellation Evaluation begins`
                   `Updating the AME database.`

4. A red cuboid will be drawn in the four viewports.
5. The views of the cuboid are 'out of alignment', and the ZOOM-C command will be used be 'line them up', so enter **ZOOM** then **C** at the prompt line:

AutoCAD prompts   `Centre point`
enter   **50,40,20<R>**
AutoCAD prompts   `Magnification or Height`
enter   **1XP<R>**

6. Repeat the above using the same centre point and scale factor in the other three viewports. Return the lower left viewport as active.
7. From the screen menu, select **MODEL**
   INQUIRY
   SOLLIST

AutoCAD prompts   `Edge/Face/Tree/<Object>`
enter   **<RETURN>** – for objects
AutoCAD prompts   `Select Objects`

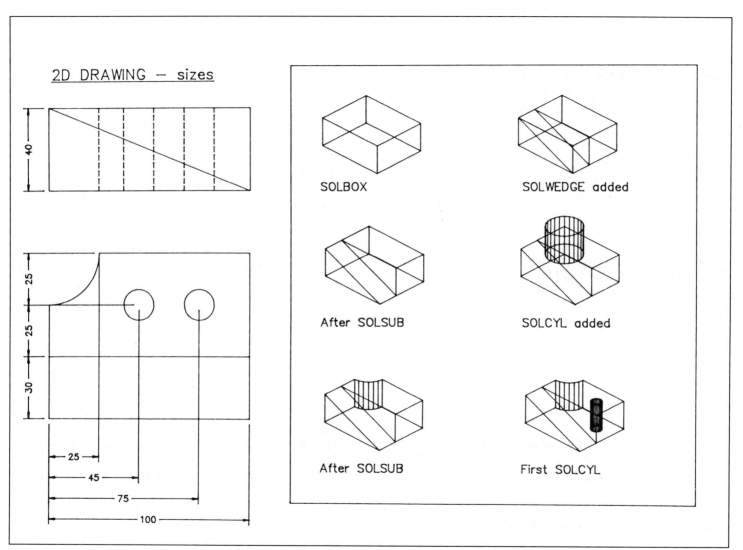

Fig. 3.1. Positioning block: construction.

respond   **pick the cuboid\<R\>**
AutoCAD prompts   `Object type=BOX (100,80,40)`
                                   `Handle=?`
                           . . . .

8. From the menu bar, select **Model**
                                   Primitives...
                                   pick Wedge then OK

AutoCAD prompts   `Baseplane/<Corner of wedge>`
                                   `<0,0,0>`
enter   **100,0,40\<R\>**
AutoCAD prompts   `Length/<Other corner>`
enter   **L\<R\>**
AutoCAD prompts   `Length` and enter **–100\<R\>**
                                   `Width` and enter **30\<R\>**
                                   `Height` and enter **–40\<R\>**

9. A red wedge will be drawn on the side of the cuboid.
10. From the screen menu, select **EDIT**
                                   CHPROP

AutoCAD prompts   `Select objects`
respond   **pick the wedge\<R\>**
AutoCAD prompts   `Change what property`
respond   **pick Colour then green\<R\>**
11. At the prompt line
enter   **SOLSUB\<R\>**
AutoCAD prompts   `Source objects...`
                                   `Select objects`
respond   **pick the red cuboid\<R\>**
AutoCAD prompts   `1 solid selected`
                                   `Objects to subtract from`
                                   `them...`
                                   `Select objects`
respond   **pick the green wedge\<R\>**
AutoCAD prompts   `Updating database.....`

12. The cuboid will be drawn with the wedge having been subtracted from it.
13. Now enter **SOLLIST** at the prompt line, and pick the composite.
AutoCAD prompts   `Object type=`
                                   `SUBTRACTION` . . . .
14. Enter **SOLCYL** at the prompt line.
AutoCAD prompts   `Baseplane/Elliptical/<Centre`
                                   `point> <0,0,0>`
enter   **0,80,0\<R\>**
AutoCAD prompts   `Diameter/<Radius>`
enter   **D\<R\>**
AutoCAD prompts   `Diameter` and enter **50\<R\>**
AutoCAD prompts   `Centre of other end/<Height>`
enter   **40\<R\>**
15. A cylinder will be drawn at the 'rear' of the composite.
16. Use **CHPROP** to change the colour of this cylinder to blue.
17. At the prompt line, enter **SOLSUB\<R\>** and
   (a) pick the composite as the source then **\<R\>**
   (b) pick the cylinder to subtract then **\<R\>**.
18. Use the **SOLCYL** command again with:
   (a) basepoint at **75,55,0\<R\>**
   (b) diameter of **15\<R\>**
   (c) height of **40\<R\>**.
19. Use **CHPROP** to change the colour of this cylinder to yellow.
20. We will now copy this cylinder to another point, so enter **COPY**
   (a) use OSNAP CENTRE and pick the top 'circle' of the cylinder
   (b) enter **@–30,0\<R\>** as the displacement.

21. Now subtract these two cylinders from the composite using **SOLSUB** with:
   (a) pick the composite as the source then**<R>**
   (b) pick the two cylinders to subtract then**<R>**.
22. SAVE your drawing – the name should be **SOLID\BLOCK** – it will be used in the next exercise.

   This is your first complete solid model.

23. Before leaving AutoCAD, enter:
   (a) SOLMESH**<R>** and pick the composite
   (b) HIDE**<R>**
   (c) SHADE**<R>**.
24. You should be quite impressed with the coloured shaded model of your first component.
25. Return the composite to wire-frame representation with:
   (a) REGEN**<R>**
   (b) SOLWIRE**<R>** and pick the composite.

❑ *Summary of commands*

This first exercise has introduced the user to solid modelling commands. A brief description of those used will now be given.

SOLBOX, SOLWEDGE, SOLCYL

These are the three of the basic solid model primitives that are available to the user. They will be discussed in more detail in a later chapter.

SOLSUB

Is the command used to subtract one solid from another. This command and others will be discussed later.

SOLLIST

This is a specialist listing command that can be used in addition to the regular LIST command.
   LIST – gives a solid's space, handle, location, etc.
   SOLLIST – details the solid's name, dimensions, handle, surface area, material, representation, etc.

SOLMESH

Applies a PFACE entity to the surfaces of a solid, and is used before the HIDE command.

SHADE

Adds shading to a solid meshed component. It should not be confused with 'rendering'.

SOLWIRE

Converts a meshed solid back to wire-frame representation.

SOLWDENS

This was used when making the standard sheet. It determines the wire density in wire-frame representations in solids. The value of the variable can be between 1 and 12, and we selected 6. If the value of SOLWDENS is increased, then the following also increase:

• the number of tessellation (curves) lines in wire-frame solids
• the number of PFACE sections in the solid
• the time taken to mesh a solid
• the time taken to perform operations
• the amount of RAM used in the display.

Notes

1. In this exercise certain words have been used which may be new to the reader, e.g. primitive, composite, etc. These words are common in solid modelling, and in the concept of this book mean:

   - primitive – any box, cylinder, wedge, extrusion, etc.
   - composite – the result of Boolean operations on two or more primitives
   - solid – general name for a composite.

2. Creating a solid model composite may require several hours of work, and it is important that the user realises that regular saving of drawings is strongly advised. This is a very worthwhile safeguard against unforeseen accidents, and I would recommend that *the user saves every 15/20 minutes*.

3. Figure 3.2 shows the solid composite at the end of the sequence of operations. Figure 3.2 has been plotted with the layer VPBORDER frozen.

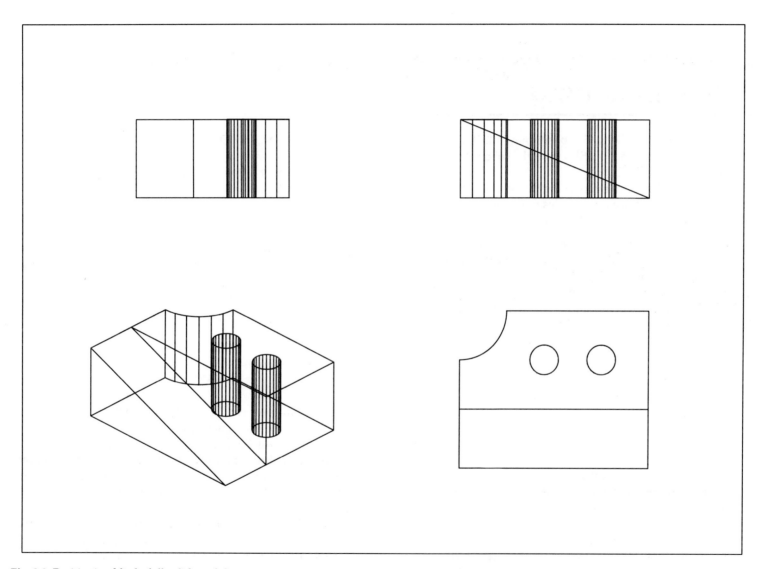

Fig. 3.2. Positioning block: full solid model.

4. Exercise 2: positioning block – profile and dimensioning

This exercise continues with the positioning block completed in Exercise 1. It will extract profiles from the solid which will be suitable for dimensioning.

Note that I will refer to layers using my handles, e.g. DIM-7, PV-9, etc. You can replace my handles with your own handles if they are different.

1. Begin a new drawing **\SOLID\BLOCK_A = \SOLID \BLOCK**. This will allow us to work on the positioning block while keeping the original drawing safe.
2. The drawing should be loaded in model space with the lower left viewport active and SOLIDS as the current layer.
3. From the screen menu select **MODEL**
   DISPLAY
   SOLPROF

AutoCAD prompts   Select objects
respond   **pick the composite<R>**
AutoCAD prompts   Updating object ... 1 solid
                  selected
then   Display hidden profile lines
                  on separate layers
enter   **Y<R>**
AutoCAD prompts   Project profile lines onto a plane
enter   **Y<R>**
AutoCAD prompts   Delete tangential edges
enter   **Y<R>**

4. The model will change colour (green) with hidden line detail added (colour 9) – think about your layers!

5. Repeat the **SOLPROF** command in the other three viewports, answering **Y** to the three prompts.
6. Make the upper left viewport active.
7. At the prompt line enter **LAYER<R>**

AutoCAD prompts   ?/Make/Set..............
enter   **S<R>**
AutoCAD prompts   New current layer
enter   **DIM-8<R>** – handle number
                  required?
AutoCAD prompts   ?/Make/Set..............
enter   **F<R>**
AutoCAD prompts   Layer name(s) to freeze
enter   **SOLIDS<R>**
AutoCAD prompts   ?/Make/Set
enter   **<RETURN>**

8. The current layer will be displayed as DIM-8 (magenta) in the status line and green profile lines and rust-coloured hidden lines will be displayed in the viewports.
9. At this stage study the views in all four viewports, and note where there are hidden (rust) and full (green) lines coincident with each other. We are going to remove these hidden lines.
10. Repeat the **LAYER** command, and freeze all visible profile lines by enter **F<R>** at the first prompt, and then **PV-*<R>** at the Layer name(s) to freeze prompt.
11. Your drawing should now display hidden lines only.

12. While still in the top left viewport, enter **EXPLODE** and pick the hidden line block – it should turn black? Now use ERASE and REDRAW to remove unwanted hidden lines.
13. Repeat step 12 in the other three viewports – in the lower right viewport the complete hidden block can be erased – think about it!
14. Now use the **LAYER** command to thaw the profile layers which were frozen in step 10, i.e. enter **T<R>** to the first prompt and **PV-*<R>** to the layer name prompt.
15. Your display should now show green profile outlines with the correct hidden line detail.
16. In the top left viewport, restore the **UCS FRONT**.
17. Use the **DIM** command to dimension this view – dimensions added should not appear in the other viewports as the dimension layers are frozen.
18. In the top right viewport, make DIM-7 (?) the current layer and restore **UCS LEFT**. Add appropriate dimensions to this view.
19. Dimension the bottom right viewport in a similar manner. The UCS for this is WCS(?)
20. Enter paper space and add any text to your border sheet.
21. Your final drawing will be similar to Fig. 4.1 (see page 20) and is now ready for plotting. Figure 4.1 has been plotted with layer VPBORDER frozen.

❏ *Summary*

This exercise introduced the user to adding dimensions to a component in separate viewports using the viewport layer concept. It also introduced the **SOLPROF** command which allows profiles of solid models to be extracted in each viewport. This command will be used in later exercises and a full explanation will be given at that stage.

The first two exercises have used commands and terms without much explanation. My idea was to introduce the user to solid modelling drawing straight away, and to make you realise that it is not as complicated as you may think. We will now investigate the main building blocks of solid modelling, i.e. the basic primitives.

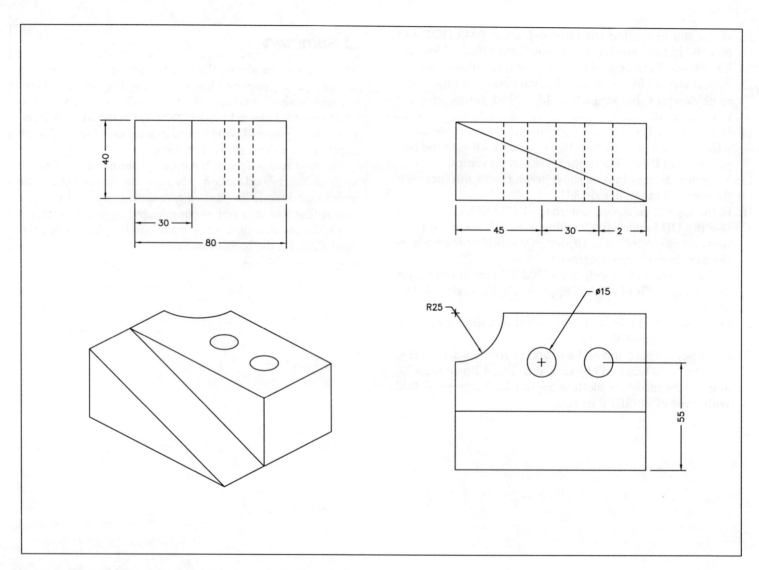

Fig. 4.1. Positioning block: (a) profile extraction and (b) dimensioning.

5. Exercise 3: the BASIC primitives

AutoCAD contains six basic primitives for solid modelling. These primitives are the BOX, WEDGE, CYLINDER, CONE, SPHERE and TORUS. We have used some of them in our previous exercises, but we will now investigate each one in detail.

There are three methods of activating the basic primitives:
  (a) keyboard entry – enter the command directly, e.g.
      SOLBOX<R>
  (b) screen menu selection: **MODEL**
                             PRISMS
                             SOLBOX
  (c) menu bar selection: **Model**
                          Primitives...
                          Dialogue box – pick as required.

It is the reader's own preference as to which method is adopted for selecting the primitives, but personally I enter the required command from the keyboard – it is the quickest way!

The BOX primitive – Fig. 5.1

1. Begin a new drawing **\SOLID\BOXPR=\SOLID\SOLA3**. Your standard sheet will be loaded with the lower left viewport active, and SOLIDS the current layer?
2. At the command prompt enter **SOLBOX<R>**.
   AutoCAD prompts `Baseplane/Centre/<Corner of box> <0,0,0>`
   enter **0,0,0<R>**
   AutoCAD prompts `Cube/Length/<Other corner>`
   enter **C<R>** – cube option
   AutoCAD prompts `Length`
   enter **100<R>**
3. A red cube will be drawn at the edge of your viewports.
4. Repeat the SOLBOX command and enter **100,0,0<R>**
   AutoCAD prompts `Cube/Length/<Other corner>`
   enter **L<R>** – length option
   AutoCAD prompts `Length` and enter 80**<R>**
   AutoCAD prompts `Width` and enter 50**<R>**
   AutoCAD prompts `Height` and enter 40**<R>**
5. Now use **EDIT** and **CHPROP** to change the colour of this object to green.
6. The two objects drawn are not 'centred' so use **ZOOM** then the centre option (**C**) to centre each viewport with **75,20,50** as the centre point and **0.5XP** as the scaling factor.
7. Enter SOLBOX again, with **0,0,0** as the corner point.
   AutoCAD prompts `Cube/Length/<Other corner>`
   enter **@60,–60,70<R>** – other corner option
8. Use **CHPROP** and change the colour of this cuboid to yellow.
9. Repeat the SOLBOX command and:
   AutoCAD prompts `Baseplane.............`
   respond **OSNAP INT and pick the left corner on top surface of the red cuboid**
   AutoCAD prompts `Cube/......`
   enter **L<R>**

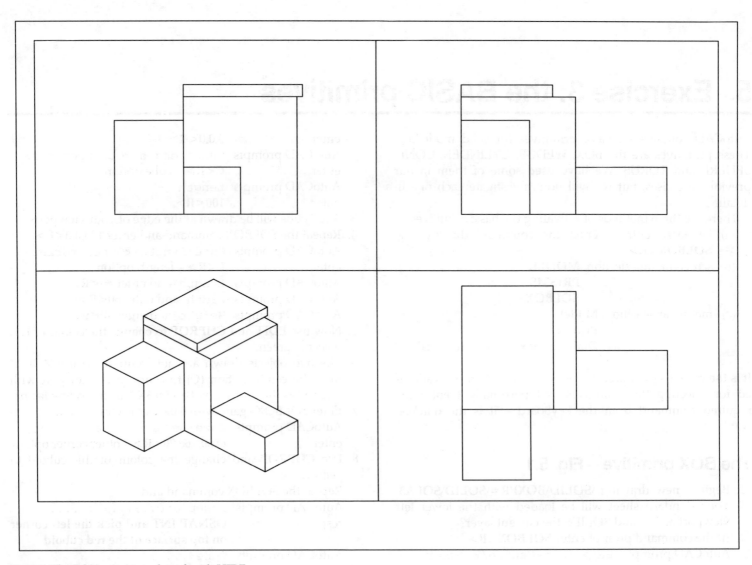

Fig. 5.1. The BOX primitive plotted with HIDE.

AutoCAD prompts `Length` and enter 80**<R>**
        `Width` and enter 105**<R>**
        `Height` and enter 10**<R>**

10. Now change this cuboid to blue.
11. Use **LIST** and **SOLLIST** on any of cuboids, noting the screen output.
12. We will now attempt to add some 'shading' to our drawing to give it more of an 'appearance'.
13. Enter **SOLMESH<R>** and pick the four cuboids. Auto-CAD will then 'mesh' these objects.
    Now enter **HIDE<R>** and then **SHADE<R>**. The result should be quite impressive.
14. Return to wire-frame representation with **REGEN** then **SOLWIRE<R>**again picking the four cuboids.
15. Enter paper space and modify your sheet layout as required.
16. Your drawing is ready for plotting.
17. Save if required.

The WEDGE primitive – Fig. 5.2

1. Begin a new drawing: **\SOLID\WEDGEPR = \SOLID\SOLA3**.
2. Enter **SOLWEDGE<R>** at the command line.
   AutoCAD prompts `Baseplane/<Corner of Wedge>`
   enter   **0,0,0<R>**
   AutoCAD prompts `Length/<Other corner>`
   enter   **L<R>** – the length option
   AutoCAD prompts `Length` and enter 100**<R>**
        `Width` and enter 80**<R>**
        `Height` and enter 60**<R>**
3. Repeat the SOLWEDGE command at **0,0,0** using the length option with a length of 40, a width of –10 and a height of 40.

4. Use **CHPROP** to change the colour of this wedge to yellow.
5. Now 'centre' the wedges with **ZOOM C**, the centre point being **50,50,50** in all viewports with a factor of **0.5XP** in the 3D viewport and **0.75XP** in the other three.
6. Use SOLWEDGE again at 0,0,0 with a length of –50, a width of 100 and a height of 80. Change the colour of this wedge to green.
7. Create another wedge at **100,80,0** and:
   AutoCAD prompts `Length/<Other corner>`
   enter   **@60,60<R>** – other corner option
   AutoCAD prompts `Height` and enter 50**<R>**
8. Now change the colour to blue, and **ROTATE** this wedge about the point **100,80,0** by **–90** degrees.
9. The last wedge to be created starts at **100,0,0**. Use the length option with length –50, width 60 and height –30. The colour is to be magenta.
10. Now **MOVE** this magenta wedge from one of its vertices, the displacement being **@0,0,30**.
11. **SOLLIST** any of the wedges.
12. Use **SOLMESH<R>** picking all wedges (crossing?) then **HIDE** and **SHADE**. Again you should be quite impressed by the result.
13. Return to wire-frame representation with REGEN and SOLWIRE.
14. Paper space to update your drawing sheet then save and plot.

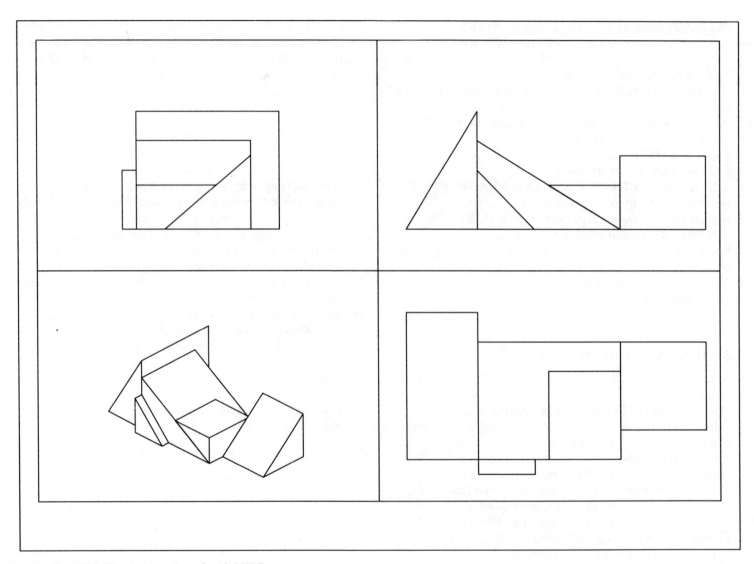

Fig. 5.2. The WEDGE primitive plotted with HIDE.

The CYLINDER primitive – Fig. 5.3

1. A new drawing \SOLID\CYLPR=\SOLID\SOLA3.
2. Enter **SOLCYL<R>**

AutoCAD prompts	Baseplane/Elliptical/ <Centre point> <0,0,0>
enter	**0,0,0<R>**
AutoCAD prompts	Diameter/<Radius>
enter	**D<R>** – diameter option
AutoCAD prompts	Diameter and enter 50**<R>**
AutoCAD prompts	Centre of other end/<Height> and enter 80**<R>**

3. Create another cylinder with SOLCYL at **40,0,0** with **15 radius** and a **height of 60**.
4. Use CHPROP to change this cylinder's colour to yellow.
5. Now enter **ARRAY** and pick the yellow cylinder. Polar array **3** items about the point **0,0**
6. Centre each viewport with **ZOOM C** about **0,0,80** with **0.5XP**.
7. Now create two cylinders using the following information:

Start point	radius	height	colour
0,0,80	20	50	green
0,0,130	50	20	blue

8. Use SOLCYL again at a centre point of **25,0,70 radius 10**.

AutoCAD prompts	Centre of other end/<Height>
enter	**C<R>** – other end option
AutoCAD prompts	Centre of other end
enter	**@50<0<0<R>**

9. Change this last cylinder to colour cyan.
10. Repeat the centre option for a cylinder at **0,–50,160** with a **radius of 10**, the other end being at **@0,20,0**
11. This last cylinder is to be magenta, and polar arrayed about the *centre of the top surface of the blue cylinder* – five items.
12. SOLLIST any cylinder, then SOLMESH, HIDE and SHADE.
13. Return to wire-frame representation with REGEN and SOLWIRE.
14. Paper space, modify, save, plot.

The CONE primitive – Fig. 5.4

1. New drawing \SOLID\CONEPR=\SOLID\SOLA3.
2. Enter **SOLCONE<R>**.

AutoCAD prompts	Baseplane/Elliptical/<Centre point> <0,0,0>
enter	**0,0,0<R>**
AutoCAD prompts	Diameter/<Radius>
enter	**50<R>** – radius option
AutoCAD prompts	Apex/<Height>
enter	**60<R>** – height option

3. Repeat the SOLCONE command the centre point being **0,0,0**

AutoCAD prompts	Diameter/<radius>
enter	**D<R>** – diameter option
AutoCAD prompts	Diameter and enter 160**<R>**
AutoCAD prompts	Apex/<Height>
enter	**–80<R>** for the height

4. Use **CHPROP** and change the colour of this cone to green.
5. Centre the viewports with ZOOM C about **0,0,0** with **0.5XP**
6. Enter SOLCONE with a centre point of **80,0,0** and **radius 30**

AutoCAD prompts	Apex/<Height>
enter	**A<R>** – apex option
AutoCAD prompts	Apex
enter	**@50<0<0<R>**

7. Change this cone to colour yellow, and the polar array it for five items about the point **0,0**.
8. Create another cone at the point **0,0,80** with a **diameter of 40**. Use the Apex option and enter **@100<180<0**. This cone should be coloured blue.
9. The last cone will be created at the point **65,0,0**. It should have a **radius of 10** and a **height of 20**. Its colour is to be magenta, and it is to be polar arrayed eight times about **0,0**.
10. SOLLIST any cone then SOLMESH, HIDE and SHADE.
11. REGEN and SOLWIRE – you should know why?
12. Paper space, etc.

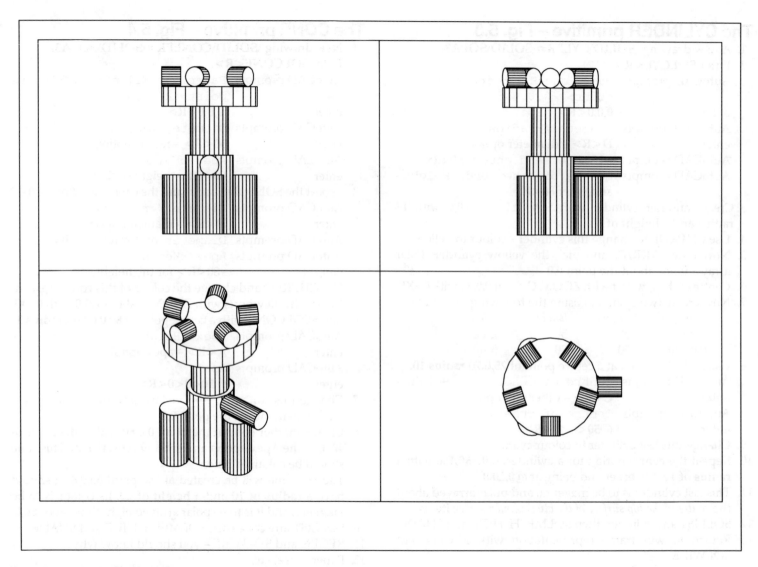

Fig. 5.3. The CYLINDER primitive plotted with HIDE.

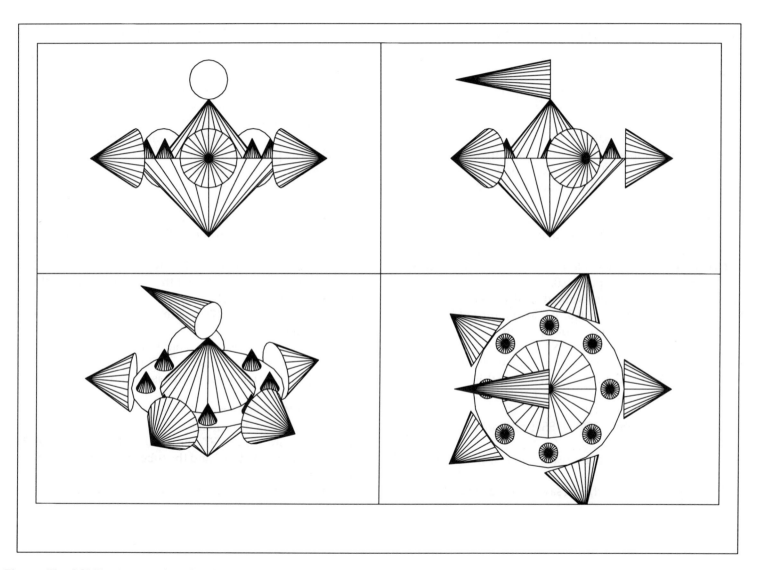

Fig. 5.4. The CONE primitive plotted with HIDE.

The SPHERE primitive – Fig. 5.5

1. Another new drawing \SOLID\SPHEREPR = \SOLID\ SOLA3.
2. Enter **SOLSPHERE<R>** at the prompt line
   AutoCAD prompts `Baseplane/<Centre of sphere>`
   `<0,0,0>`
   enter **0,0,0<R>**
   AutoCAD prompts `Diameter/<Radius of sphere>`
   enter **50<R>** – radius option
3. Repeat the SOLSPHERE command, but enter **70,0,0** as centre
   AutoCAD prompts: Diameter/<Radius of sphere>
   enter **D<R>** – diameter option
   AutoCAD prompts `Diameter` and enter 40**<R>**
4. **CHPROP** this sphere's colour to yellow, and polar array it three times about the point 0,0.
5. Now centre the viewports about **0,0,0** at **0.5XP**.
6. Create another sphere at **0,0,70** with a **radius of 20**.
7. CHPROP for this sphere – colour green.
8. Restore the UCS named FRONT.
9. Array (polar) the green sphere about **0,0** – three items.
10. Return to WCS and create a sphere at **42,0,56** of **radius 20**.
11. This sphere should be blue, and polar arrayed for three items about the point 0,0.
12. SOLLIST any sphere, SOLMESH all spheres, HIDE and SHADE.
13. Return to wire-frame representation with REGEN and SOLWIRE.
14. Paper space to add text, etc., save and plot if needed.

The TORUS primitive – Fig. 5.6

1. New drawing......................
2. At the prompt line enter **SOLTORUS<R>**
   AutoCAD prompts `Baseplane/<Centre of torus)`
   enter **0,0,0<R>**
   AutoCAD prompts `Diameter/<Radius of torus>`
   enter **50<R>** – radius option
   AutoCAD prompts `Diameter/<Radius of tube>`
   enter **20<R>**
3. Create another torus at **110,0,0** with a torus radius of 30, and a tube radius of 10.
4. CHPROP for this torus – colour green.
5. Polar array the green torus (three items) about the point 0,0.
6. Now centre the view in all viewports with ZOOM C, the centre point being **0,0,80** and the scaling factor **0.4XP**.
7. Restore the UCS FRONT.
8. Use SOLTORUS at **0,75,0** with a torus radius of 50 and a tube radius of 20.
9. Change this last torus to colour blue.
10. Create another torus (same UCS) at **0,40,110** the colour being yellow. The torus radius is 30 and the tube radius is 10.
11. Restore WCS and array this yellow torus about the point 0,0 there being three items.
12. Now restore UCS LEFT and create a torus at **0,145,0**. The torus radius is to be 65 and the tube radius 10 – colour cyan.
13. Return to WCS.
14. SOLLIST, SOLMESH, HIDE, SHADE and return to wire-frame.
15. Paper space, etc.

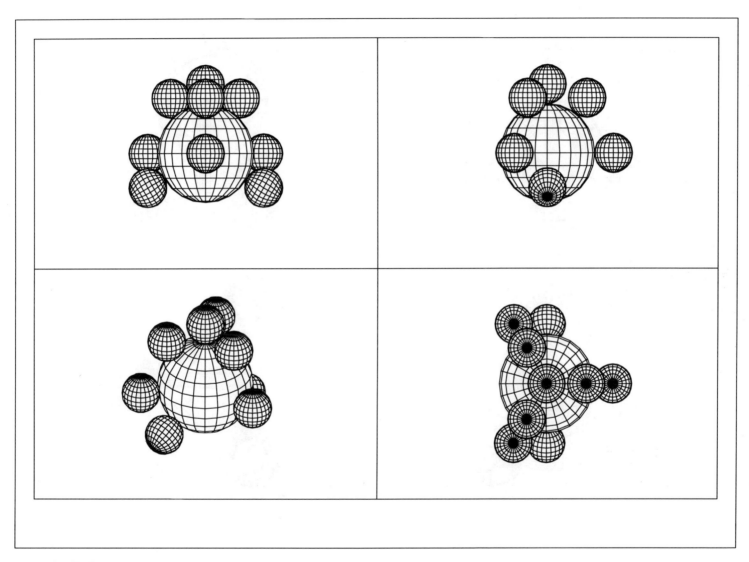

Fig. 5.5. The SPHERE primitive plotted with HIDE.

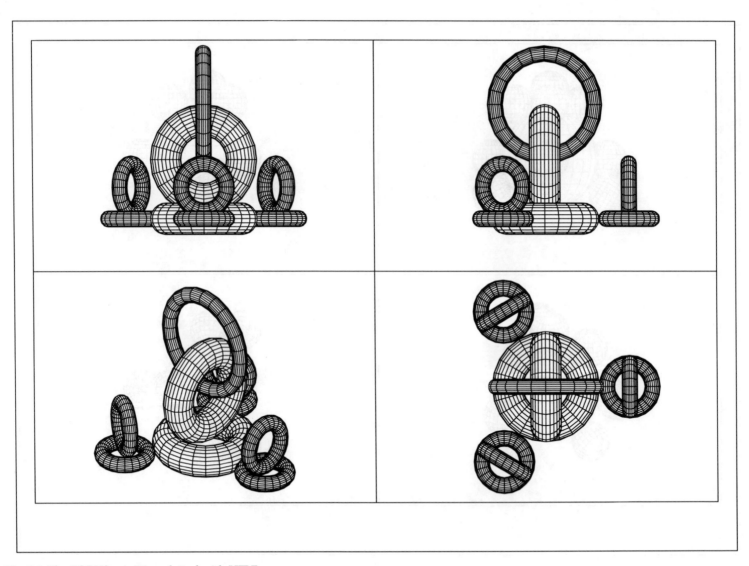

Fig. 5.6. The TORUS primitive plotted with HIDE.

This chapter has introduced the reader to the six basic primitives used with AutoCAD. I have tried to make the exercises interesting by creating a 'design' for each primitive and also tried to demonstrate the different options available in the commands.

Baseplanes and construction planes

All of the six primitive SOL commands has a prompt '**Baseplane**'. This allows the solid to be set to some orientation other than the current X–Y plane. AutoCAD allows four different types of base:

1. The WCS which is the base for all the various UCSs.
2. A UCS which can be orientated anywhere.
3. A baseplane (B) which is a temporary UCS.
4. A construction plane (CP) which is also temporary.

When B or CP is entered, a further prompt is displayed:

```
Entity/last/Zaxis/View/XY/YZ/ZX/<3point>
```

Tasks

1. Investigate some of the options we have not used in this exercise, e.g. Elliptical.
2. Investigate a torus with a tube radius of 0.
3. Produce a single drawing of all six primitives.
4. Investigate the baseplane option.

6. Exercise 4: the SWEPT primitives

The previous chapter introduced the reader to the six basic primitives. AutoCAD also allows SWEPT primitives which are of two types:

(a) extrusions
(b) revolutions.

Both these primitives are generated by sweeping an existing shape through either a straight or circular path. We will demonstrate the two swept primitives with three examples of each. Refer to Fig. 6.1 which details in 2D the basic sizes for each component to be drawn. My sizes are only for reference and you can use your own in desired, but I would advise you to try and keep the 'overall' size the same as mine. This will assist when the ZOOM C option is used.

As before, the commands can be obtained from either the keyboard, the screen menu or the pull-down menu area as follows:

1. Keyboard, enter **SOLEXT<R>**
2. Screen menu **MODEL**
   SOLEXT
3. Menu bar **Model**
   Extrusion

The extrusions

Extrusion 1 – a letter – Fig. 6.2

1. Begin a new drawing **SOLID\\SWEX1=\\SOLID\\SOLA3**.
2. Make the top right viewport active and restore UCS LEFT.

3. Use the ZOOM C option with **50,40,–25** at **0.5XP** in the 3D viewport and 50,40,–25 at **1XP** in the other three.
4. Use the PLINE command from 0,0 to draw half the letter M shape then MIRROR it about the vertical centre line.
5. Now PEDIT to make one complete polyline – if you have had to draw the shape with several polylines don't worry. The PEDIT and join option will overcome this.
6. Temporary save at this stage, the name is anything.
7. Now enter **SOLEXT<R>**
   AutoCAD prompts `Select objects`
   respond **pick the polyline letter<R>**
   AutoCAD prompts `Height of extrusion`
   enter **–50<R>**
   AutoCAD prompts `Extrusion taper angle<0>`
   enter **0<R>**
8. The letter will now be extruded as Fig. 6.2.
9. SOLMESH, HIDE and SHADE as in previous exercises, then return to wire-frame representation with REGEN and SOLWIRE.
10. In paper space alter your A3 sheet, save and plot.
11. Open the temporary drawing of the letter.
12. Repeat the SOLEXT command, entering **–50** as the height of the extrusion, but this time enter a taper angle of **30** – Fig. 6.3.
13. Save and plot this drawing if you want.

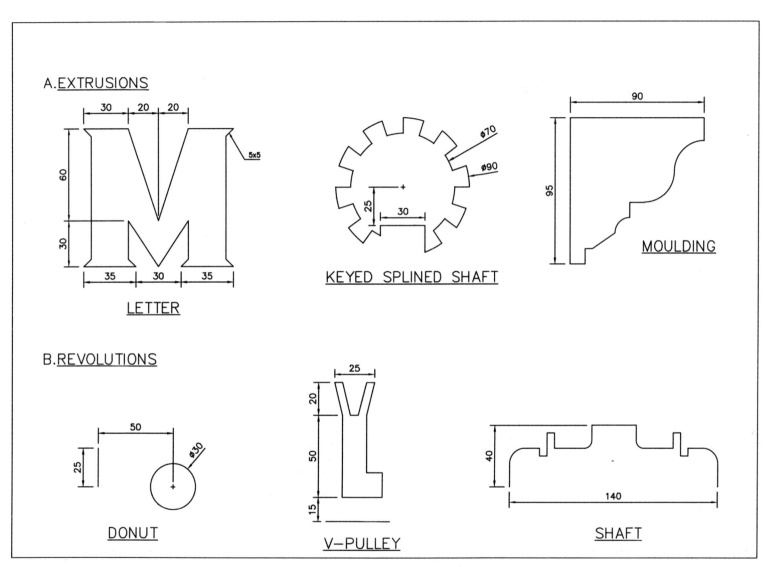

Fig. 6.1. Sizes for SWEPT primitives.

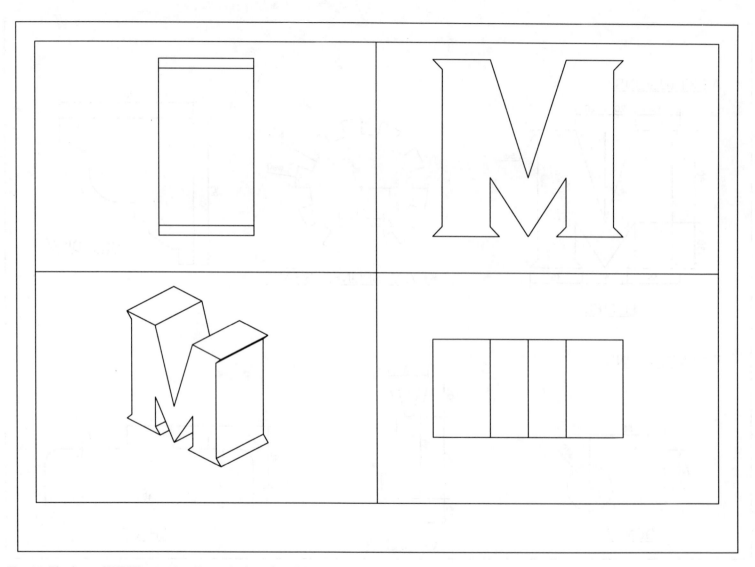

Fig. 6.2. The letter SWEPT primitive (0 taper) plotted with HIDE.

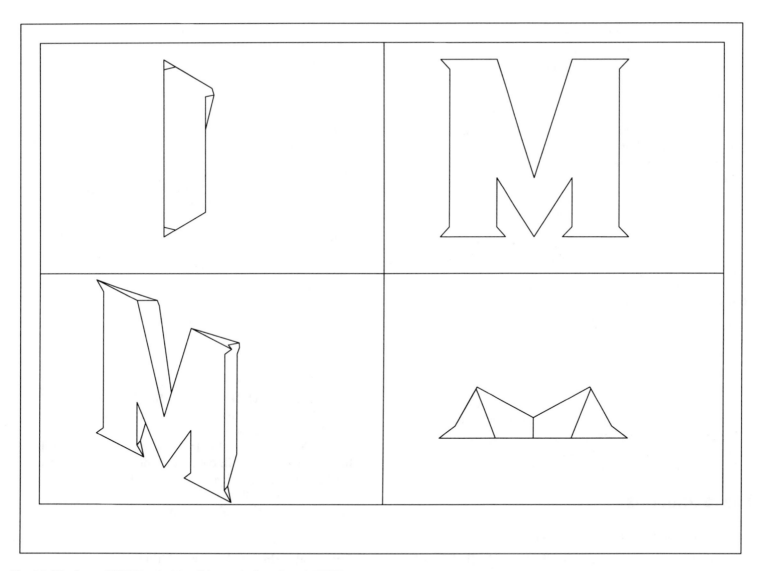

Fig. 6.3. The letter SWEPT primitive (30 taper) plotted with HIDE.

Extrusion 2 – a keyed splined shaft – Fig. 6.4

1. Begin a new drawing **\SOLID\SWEX2=\SOLID\SOLA3**.
2. With the lower right viewport active ZOOM C about **0,0,25** at **1XP** in all viewports.
3. With reference to the sizes in Fig. 6.1 draw two concentric circles at 0,0 and then array a line (20 items) about the centre point. Add the cut-out then use the TRIM command to give the final shape.
4. Using PEDIT turn the complete spline into one polyline.
5. Now enter **SOLEXT** with a height of 50 and 0 taper selecting the spline.
6. SOLMESH, HIDE, SHADE then return to wire-frame representation.
7. Investigate the taper option, but be prepared for a wait as the calculation takes some time – especially on a 386 machine.

Extrusion 3 – a moulding – Fig. 6.5

1. Begin a new drawing **\SOLID\SWEX3=\SOLID\SOLA3**.
2. Make top left viewport active and restore UCS FRONT.
3. ZOOM C using **50,50,50** at **0.5XP** in the 3D viewport, and **1XP** in the other three.
4. Use the PLINE command to draw the moulding section from 0,0. The actual 'curved edge' shape is at your discretion, but make the vertical and horizontal sizes as Fig. 6.1.
5. SOLEXT the shape a height of 100 with 0 taper.
6. SOLMESH, etc., and then paper space, etc.
7. Try a longer extrusion with a taper angle added.

The revolutions

Revolution 1 – a donut toroid? – Fig. 6.6

1. Another new drawing **\SOLID\SWEX4=\SOLID\SOLA3**.
2. Active viewport is the top left with UCS FRONT.

3. Draw a circle at 0,0 of diameter 30, and a line at –50,0 the length being @0,25.
4. ZOOM C about **–50,0** at **0.8XP** in all viewports.
5. At the prompt line

enter	**SOLREV<R>**
AutoCAD prompts	Select objects
respond	**pick the circle<R>**
AutoCAD prompts	Axis of revolution – Entity/ X/Y/<Start point>
enter	**E<R>**
AutoCAD prompts	pick entity to revolve around
respond	**pick the bottom end of the red line<R>**
AutoCAD prompts	Angle of revolution<full circle>
enter	**210<R>**

6. Erase the line.
7. SOLMESH, etc.
8. Before leaving this exercise, investigate different angle of revolution, as well as picking the top of the red line as the axis.

Revolution 2 – a V-pulley – Fig. 6.7

1. New drawing **\SOLID\SWEX5=\SOLID\SOLA3**.
2. Top left viewport active, UCS FRONT with ZOOM C at 0,0,0 and **0.5XP** in all viewports.
3. Use PLINE starting at 0,15 to draw the pulley section. If you have to use the PLINE command more than once, remember PEDIT and join all your segments. Remember to add the centre line from 0,0 for any length.
4. Now SOLREV the pulley shape for 360° about the centre line.
5. SOLMESH, etc.
6. Investigate other angles of revolution, etc.

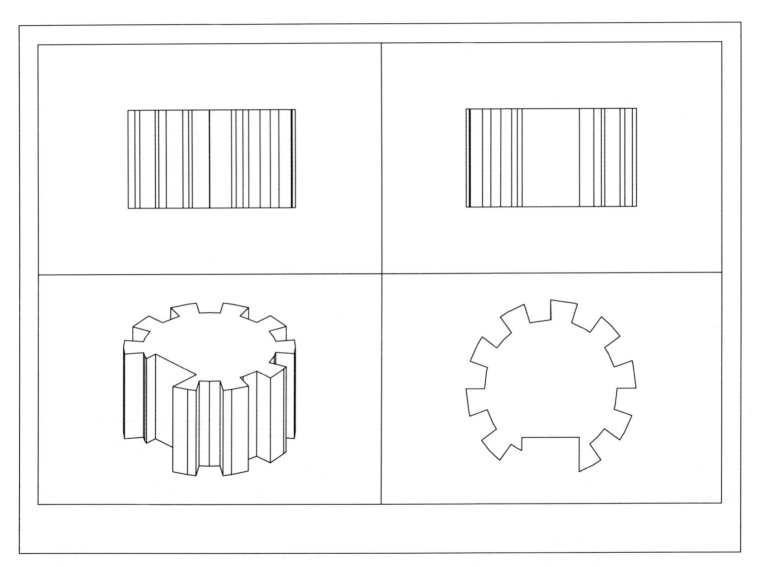

Fig. 6.4. The shaft SWEPT primitive (0 taper) plotted with HIDE.

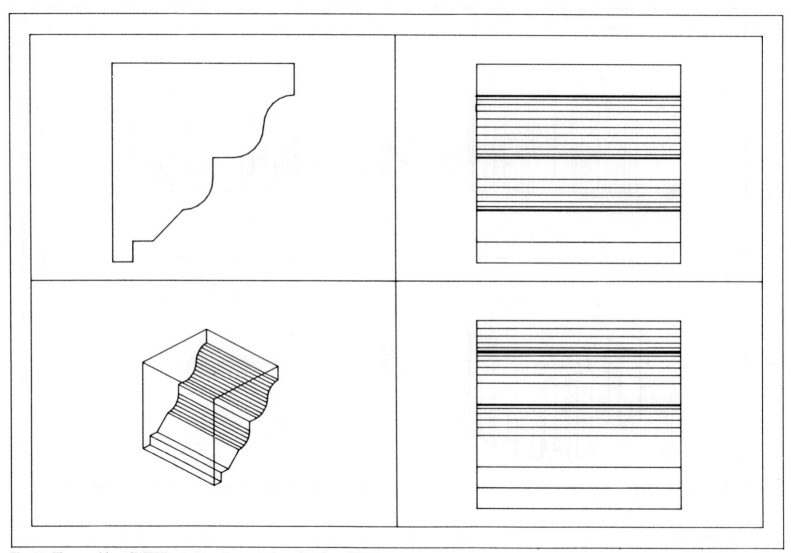

Fig. 6.5. The moulding SWEPT primitive (0 taper) plotted without HIDE.

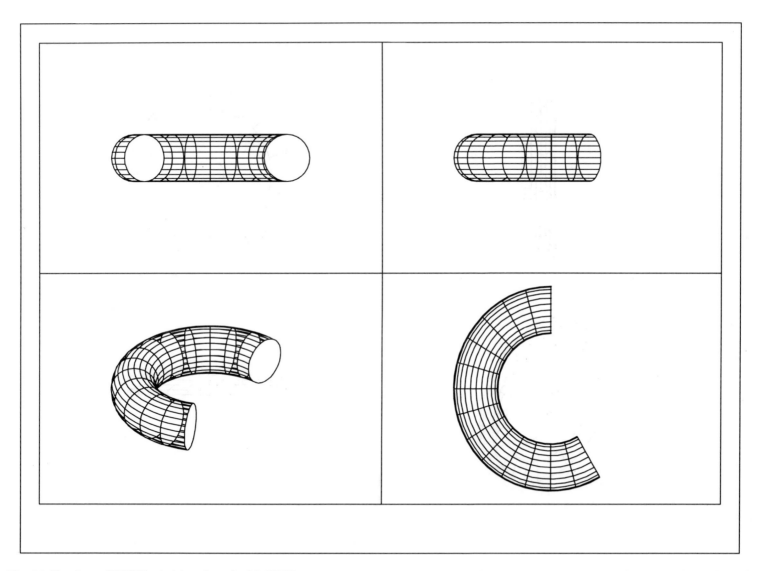

Fig. 6.6. The donut SWEPT primitive plotted with HIDE.

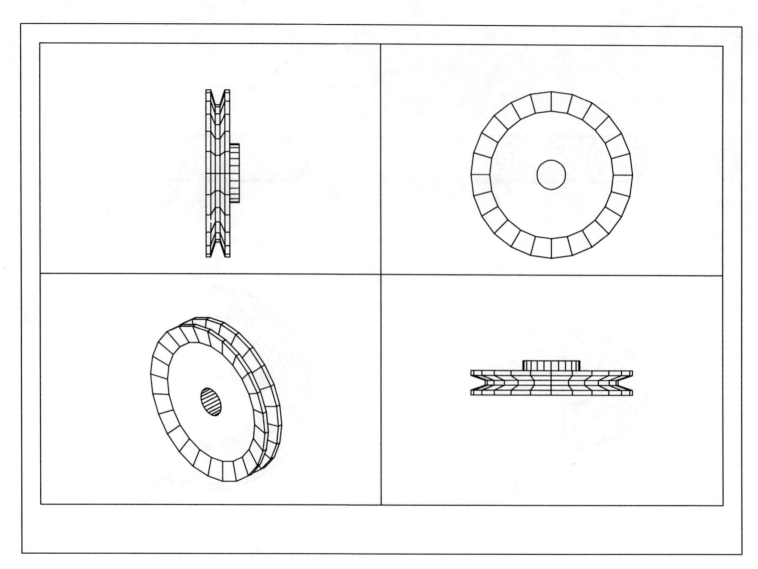

Fig. 6.7. The V-pulley SWEPT primitive plotted with HIDE.

Revolution 3 – a shaft – Fig. 6.8.

1. Last new drawing **\SOLID\SWEX6=\SOLID\SOLA3**.
2. Make the bottom right viewport active with ZOOM C about **70,0** at **0.5XP** in all viewports.
3. Use PLINE from 0,0 to draw a shaft outline consisting of line and arc segments. I drew half a section then used MIRROR and PEDIT the make one complete polyline.
4. SOLREV the polyline shape about the two shaft endpoints.
5. The rest you should know by now.

❏ *Summary of commands*

This exercise has introduced the reader to two very powerful solid modelling commands. Both are used with 'outline' shapes to extrude or revolve thus giving a solid object.

SOLEXT

Creates a solid primitive by extruding a region, circle, polyline or 3D poly entity. Very irregular shaped sections can be extruded at the one time. The extrusion is always perpendicular to the selected entities plane. The user selects

1. The object to be extruded.
2. The height of the extrusion.
3. The extrusion taper angle.

SOLREV

Produces a solid by revolving a region, circle, polyline or 3D poly about an axis. One entity can only be revolved at a time, and the user selects

1. The object to be revolved.
2. The axis of revolution.
3. The angle of revolution.

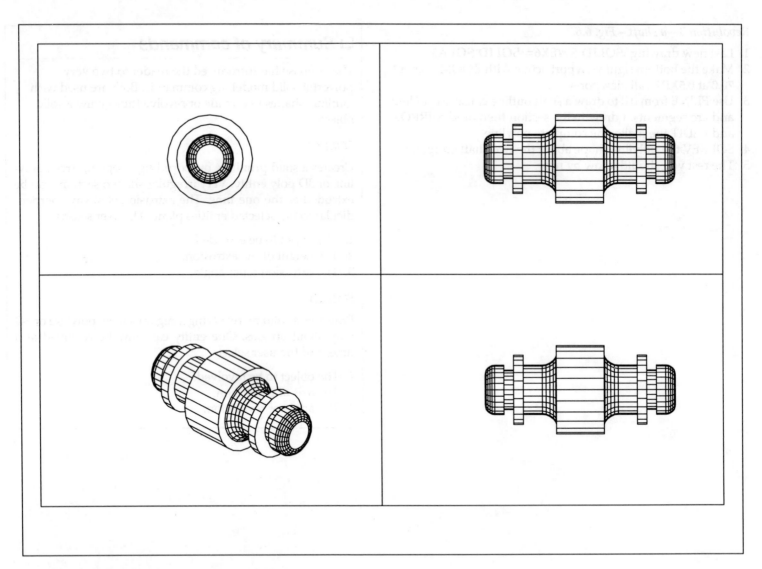

Fig. 6.8. The shaft SWEPT primitive plotted with HIDE.

7. Exercise 5: the EDGE primitives

EDGE primitives allow the user to modify the edges of existing solids, and AutoCAD allows two types:

(a) chamfered edges
(b) filleted edges.

Again these primitives will be demonstrated with a series of exercises, each of which uses our standard A3 sheet. The commands can be activated from

1. The keyboard, e.g. **SOLCHAM<R>**.
2. The screen menu with **MODEL**
   MODIFY
   SOLCHAM

3. Pull-down menu bar with **Model**
   Modify >
   Chamfer Solids

As before keyboard entry will be used.

The CHAMFERED edges

Exercise A – Fig. 7.1

1. Begin a new drawing **\SOLID\EDGEX1=\SOLID\SOLA3**.
2. With the 3D viewport active, ZOOM C at 0,0,50 with 0.75XP in the 3D viewport and 1XP in the other three.
3. Use SOLCYL to create a cylinder at 0,0,0 of diameter 100 and height 80.

4. At the prompt line enter **SOLCHAM<R>**
   AutoCAD prompts   pick base surface
   respond   **pick the circle on the top surface<R>**
   (You may have to enter N<R> until the circle is highlighted.)
   AutoCAD prompts   `pick edges of this face to be chamfered`
   respond   **pick the same edge as before<R>**
   AutoCAD prompts   `1 edge selected`
   then   enter distance along base surface
   enter   **15<R>**
   AutoCAD prompts   enter distance along adjacent surface
   enter   **15<R>**
5. Repeat the SOLCHAM command on the bottom circle and enter two different chamfer distances – I used 20 and 30.
6. Try SOLLIST on the chamfer and then on the cylinder.
7. SOLMESH, HIDE, SHADE but return to wire-frame as previous exercises.
8. Paper space to save and plot.

Exercise B – Fig. 7.2

1. New drawing **\SOLID\EDGEX2=\SOLID\SOLA3**.
2. ZOOM C at 75,60,50 with 0.6XP in the 3D viewport and 0.8XP in the other three. Make lower left viewport active.
3. Use SOLBOX to create a cuboid at 0,0,0 of length 150, width 120 and height 100.
4. Enter **SOLCHAM<R>**.

AutoCAD prompts   pick base surface
respond   **pick the bottom surface<R>**
(Remember N if it is not highlighted!)
AutoCAD prompts   pick edges of this face to be chamfered
respond   **pick *only* the rightmost bottom edge<R>**
AutoCAD prompts   enter   distance   along   base surface
enter   **20<R>**
AutoCAD prompts   enter   distance   along adjacent surface
enter   **20<R>**
5. Repeat the SOLCHAM command and:
   (a) pick the top surface**<R>** at the first prompt
   (b) pick ALL FOUR TOP EDGES**<R>** at the second prompt
   (c) enter the first chamfer distance as 15
   (d) enter the second chamfer distance as 25.
6. SOLLIST the chamfer, then SOLMESH, HIDE, etc.

Exercise C – Fig. 7.3

1. With a new drawing (the name should be obvious by now) ZOOM C about the point 75,60,50 at 0.5XP in 3D viewport and 0.8XP in the other three, and activate the lower left viewport.
2. Use SOLWEDGE to create a wedge solid at 0,0,0 with a length of 150, a width of 120 and a height of 100.
3. SOLCHAM the sloped surface and pick all four edges. The chamfer distances are both 15.

The FILLETED edges

Exercise D – Fig. 7.4

1. New drawing and create a cylinder with the sizes and zoom factor as Exercise A.
2. Enter **SOLFILL<R>**

AutoCAD prompts   pick edge of solids to be filleted
respond   **pick the top circle<R>** – remember N
AutoCAD prompts   1 edge selected
                   Diameter/<Radius> of fillet
enter   **10<R>** – radius option
3. Repeat the SOLFILL command on the bottom circle, the fillet diameter being 30.
4. SOLLIST on the fillet, then on the solid.
5. SOLMESH......

Exercise E – Fig. 7.5

1. Yet another new drawing with name **SOLID**\\...............
2. Create a box as Exercise B, i.e. same sizes and zoom effect.
3. Enter **SOLFILL<R>**
   AutoCAD prompts   pick edges of solids to be filleted
   respond   **pick the bottom left-hand (lowest) edge<R>**
   AutoCAD prompts   Diameter/<Radius> of fillet
   enter   **15<R>**
4. Repeat the SOLFILL command and fillet all four top edges with a diameter of 40.

Exercise F – Fig. 7.6

1. Guess what? – a new drawing **SOLID**\\...............
2. Create a wedge the same as Exercise C, zoom as well.
3. SOLFILL and pick the three edges of the left side. The fillet has to have a radius of 15.

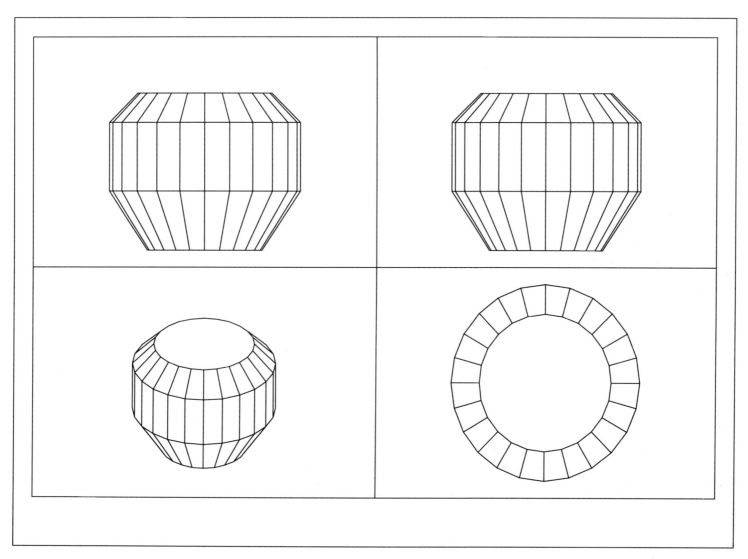

Fig. 7.1. The EDGE primitive – CHAMFER with SOLCYL with HIDE.

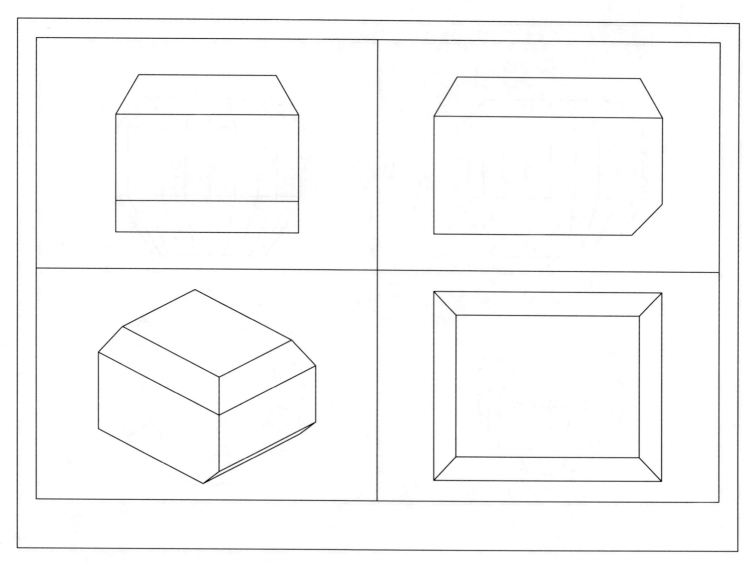

Fig. 7.2. The EDGE primitive – CHAMFER with SOLBOX with HIDE.

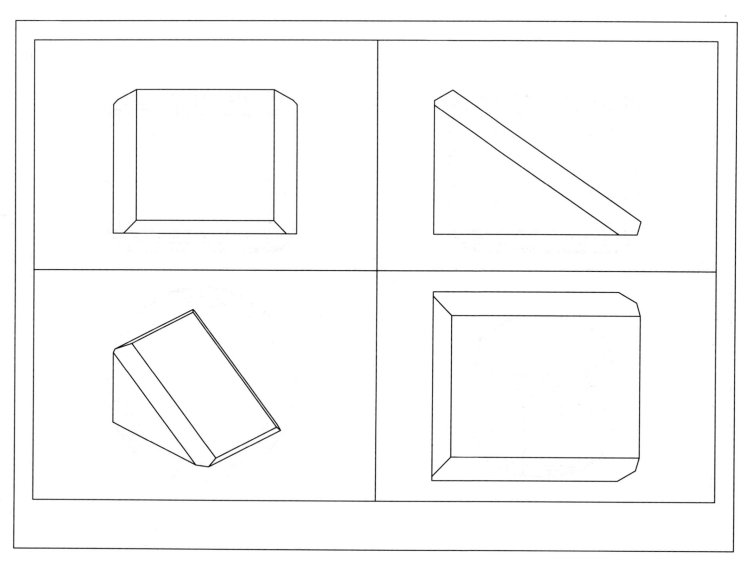

Fig. 7.3. The EDGE primitive – CHAMFER with SOLWEDGE with HIDE.

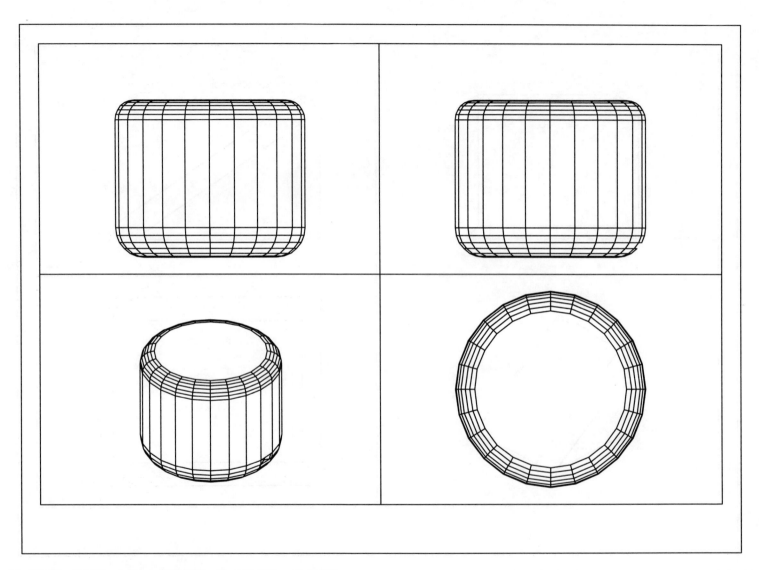

Fig. 7.4. The EDGE primitive – FILLET with SOLCYL with HIDE.

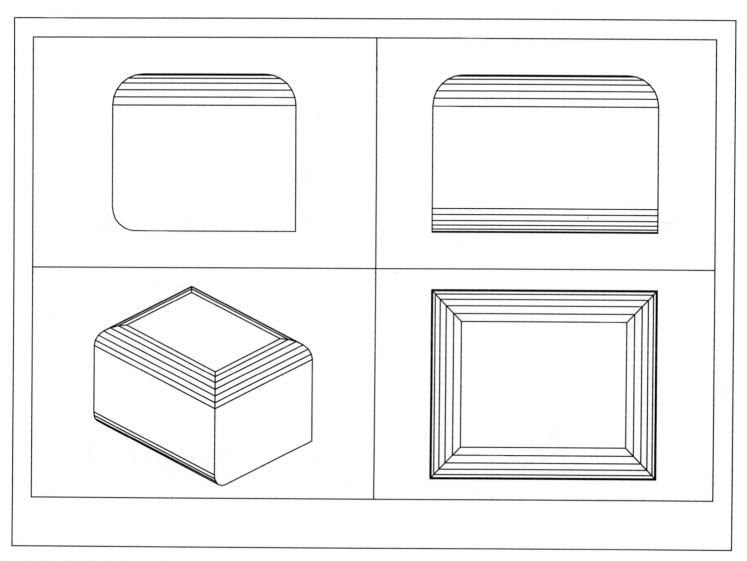

Fig. 7.5. The EDGE primitive – FILLET with SOLBOX with HIDE.

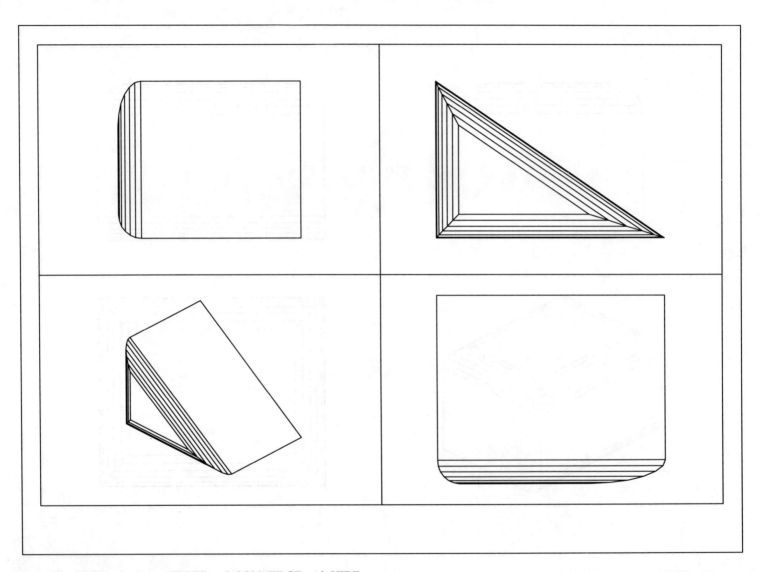

Fig. 7.6. The EDGE primitive – FILLET with SOLWEDGE with HIDE.

❏ *Summary of commands*

The two commands for the edge primitives are quite simple to understand:

- **SOLCHAM** – creates a bevelled edge on a solid. It is a separate solid that is subtracted from an existing solid.
- **SOLFILL** – creates a filleted edge and is a separate solid subtracted from an existing solid.

Tasks

1. Refer to Fig. 7.7.
2. Create a box at 0,0,0 that is 150 in length, 120 wide and 100 high.
3. ZOOM C as in Exercise B.
4. Create a cylinder at 75,60,0 of radius 40 and 100 height.
5. Subtract the cylinder from the box (SOLSUB)
6. SOLFILL the bottom of the box with radius 15.
7. SOLCHAM the top of the box with distances 15.
8. SOLFILL the top of the cylinder with radius 15.
9. SOLMESH, HIDE, etc.
10. Interesting result, especially at the top surface.

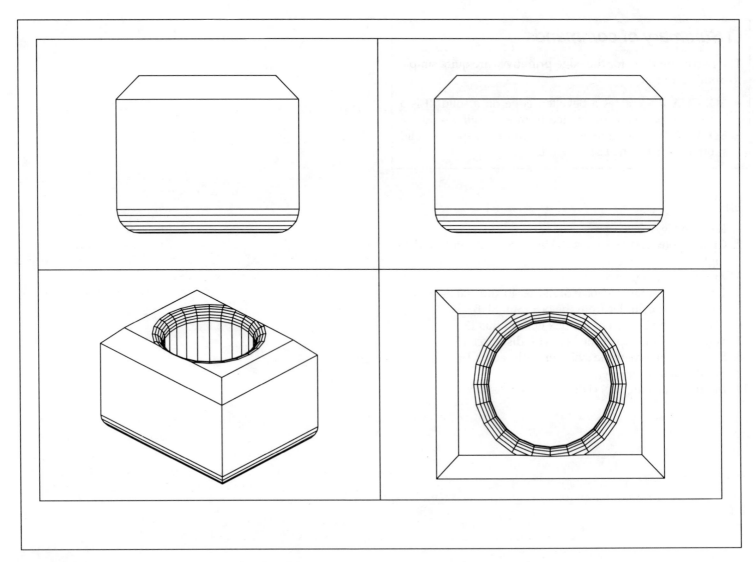

Fig. 7.7. EDGE primitive drawing task plotted with HIDE.

8. Boolean operations

The primitive solids that we have used are the 'basic tools of solid modelling' and are used to create composite solids – so-called because they are composed of two or more solids or primitives, i.e. a box is a primitive solid, but a box with a cylinder subtracted from it is a composite solid. Composite solids are also called *Boolean solids* as they are created after Boolean operations have been performed on them, a process which is called *constructive solid geometry* or CSG.

The Boolean operations which AutoCAD can perform are :

(a) union
(b) difference
(c) intersection.

The union operation

This operation involves joining two or more solids to form a single composite solid. In Boolean arithmetic, a union of two sets results in a single set containing every element (or member) that was contained in the original sets. For solids, a union includes every part of each of the original solids, and I have illustrated this by a welded type joint shown in Fig. 8.1(a), which also gives the Venn diagram.

SOLUNION is the command for a union operation, and the solids selected can be:

1. Overlapping – having a common volume.
2. Adjacent – no common volume but 'touching'
3. Non-adjacent – with a gap between the solids.

The difference operation

This is also called subtraction and is the process of removing one or more solids from another solid thereby creating the composite. In Boolean arithmetic, a difference of two sets results in the single set which contains only those members in the original set that do not belong to the 'subtracted' set. With solids, subtraction includes only the volume of the first solid that is not in the second solid, and I have illustrated this with a drilled hole as shown in Fig. 8.1(b).

SOLSUB is the AutoCAD command for the subtraction operation, and the user selects:

1. The source solid.
2. The solids to be subtracted from the source.

The intersection operation

This process forms a composite solid from solids which have a common volume. In Boolean arithmetic, the intersection of two sets results in a single set having those elements which are common to the original sets. For solids, intersection results in a composite having the same common volume as the primitives, and I have illustrated this in Fig. 8.1(c) by a 'hole' – if such an idea is possible.

SOLINT performs the intersection operation, the user selecting the solids to be intersected.

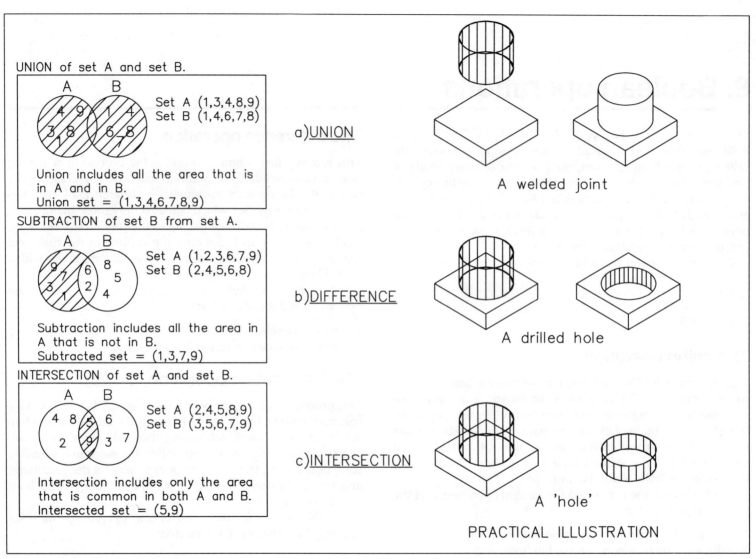

Fig. 8.1. The Boolean operations.

Separation of solids

The union, subtraction and intersection operations result in the primitives being 'joined' into a composite solid. These composites can also be taken apart or separated. Separation undoes the last Boolean operation, and it is possible to separate solids until the basic primitives are obtained.

SOLSEP is the command for solid separation, the user simply selecting the composite.

Cutting solids

Solids (primitives and composites) can be cut by a 'cutting plane' for several reasons, e.g. displaying a certain face or to obtain a true shape, etc.

SOLCUT is the command to set a cutting piane to form two separate solids or to cut away part of the existing solid. The cut is always flat along some plane.

The CSG tree

When a composite solid has been made, it can be thought of as being assembled according to its own constructive solid geometry or CSG tree. Every composite that is made is dependent on:

1. The primitives included in its construction.
2. The Boolean operations performed.
3. The order of these Boolean operations.

The CSG tree is a complete record of the Boolean operations and the primitives involved and the order of the operations. The tree is considered upside-down, the final composite being at the top, and the parts making the composite are at the bottom.

Figure 8.2 illustrates two possible CSG trees for a composite made from box, wedge and cylinder primitives. In Fig. 8.2(a) the tree begins at the lowest level with a solid box and a solid cylinder. At the middle level, the cylinder is subtracted from the box, and a solid wedge is drawn. At the top level, the wedge and box with the hole are unioned. In Fig. 8.2(b) the lowest level has the solid box and solid wedge, the middle level has the box and wedge unioned while the solid cylinder is drawn, and the top level has the cylinder subtracted from the wedge/box composite.

Balanced and unbalanced trees

A CSG tree has only two branches. This branch may divide into another two branches and so on, and a complicated tree structure can result. A CSG tree is thus structured for operating in pairs. Figure 8.3(a) illustrates a balanced tree of three levels for a composite made from four primitives. The same composite could have been made from an unbalanced tree of four levels as Fig. 8.3(b).

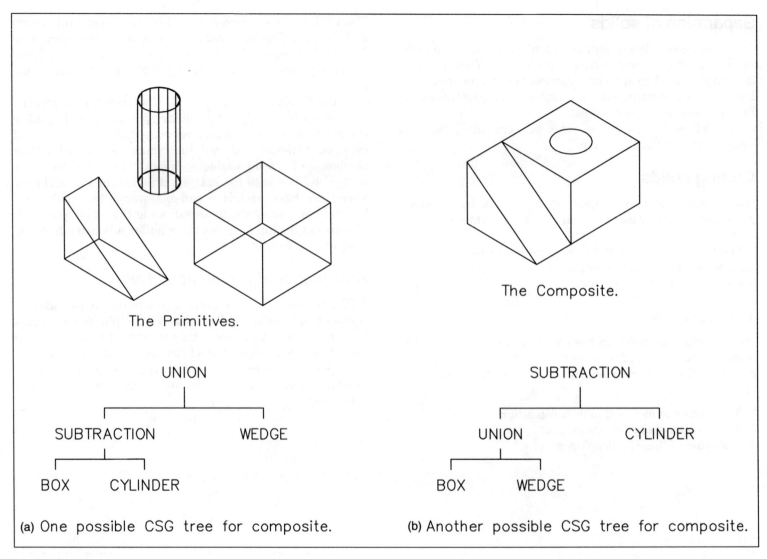

The Primitives.

The Composite.

UNION

SUBTRACTION

SUBTRACTION   WEDGE

UNION   CYLINDER

BOX   CYLINDER

BOX   WEDGE

(a) One possible CSG tree for composite.

(b) Another possible CSG tree for composite.

Fig. 8.2. CSG tree.

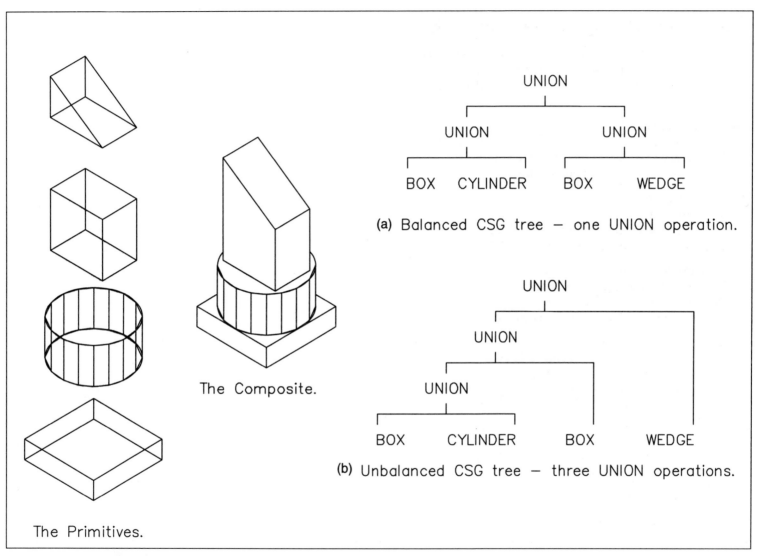

The Primitives.

The Composite.

UNION

UNION

UNION

BOX   CYLINDER   BOX   WEDGE

(a) Balanced CSG tree — one UNION operation.

UNION

UNION

UNION

BOX   CYLINDER   BOX   WEDGE

(b) Unbalanced CSG tree — three UNION operations.

Fig. 8.3. Balanced and unbalanced CSG tree using UNION operation.

Activity – predicting Boolean operations

Refer to Fig. 8.4. which illustrates four composite solids made from two/three primitives. Try and predict the following Boolean operations by sketching the expected result from the following:

(a) **Box and wedge**
1. union of the box and wedge.
2. subtraction of the wedge from the box.
3. intersection of the box and wedge.

(b) **Box and cylinder**
1. union of the box and cylinder.
2. subtraction of the box from the cylinder.
3. subtraction of the cylinder from the box.

(c) **Wedge and cylinder**
1. union of the wedge and cylinder.
2. subtraction of the cylinder from the wedge.
3. intersection of the wedge and cylinder.

(d) **Box and two cylinders**
1. subtraction of the two cylinders from the box.
2. intersection of the box and cylinders.
3. intersection of the two cylinders.

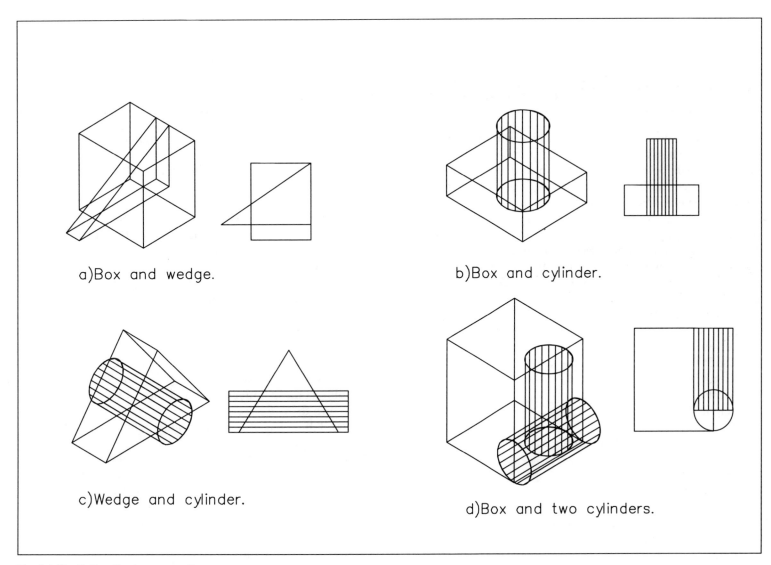

a)Box and wedge.

b)Box and cylinder.

c)Wedge and cylinder.

d)Box and two cylinders.

Fig. 8.4. Predicting Boolean operations.

9. Exercise 6: a machine support

In this exercise we will create a machine support composite made from the box, cylinder and wedge primitives. The Boolean operations will be union, subtraction and intersection. The solid will then be dimensioned using the correct viewport dimension layers. The exercise is quite simple, and you should have no problems following the steps. Try and reason out the co-ordinate input by sketching each primitive as it is drawn. The steps involved will now be simplified, as you should have a good grasp of the prompts for the various SOL commands.

1. Begin a new drawing **SOLID\MACHSUPP =\SOLID\ SOLA3**.
2. ZOOM C about 25,40,40 at 1XP in all viewports.
3. Make the lower left viewport active with SOLIDS (red) the current layer.
4. Use SOLBOX to create the base at 0,0,0 with
   length 50
   width 75
   height 15.
5. This gives Fig. 9.1.
6. Create another box on top of the base with
   start point 50,60,15
   length –50
   width –45
   height 50.
7. Use SOLCYL to create a cylinder on top of this box with
   start point 25,30,65
   diameter 50
   centre option – centre point @0,30 – Fig. 9.2.

8. Create a wedge at 0,60,65 with a length of 30, width 50 and a height of 50.
9. Rotate this wedge about the point 0,60,65 by –90° to give Fig. 9.3.
10. Intersect the wedge and cylinder with SOLINT.
11. Union the cut cylinder and the two boxes.
12. Set a new UCS on the slope face of the curved surface with:
    (a) UCS 3-point
    (b) origin at midpoint of line AB – Fig. 9.4
    (c) X axis at end of point B
    (d) Y axis at intersection of point C
    (e) UCSICON set to OR??
    (f) save this UCS as SLOPE.
13. Using this UCS (slope) create a cylinder at 0,12,0 with a radius of 8 and a height of –40 giving Fig. 9.5.
14. Subtract the cylinder from the composite.
15. Return to WCS.
16. Use SOLCYL at 10,7.5,0 to create a cylinder of diameter 10 and height 15.
17. Now array this last cylinder (a) rectangular
                                    (b) 2 rows and 2 columns
                                    (c) 60 row distance
                                    (d) 30 column distance.
18. Subtract the four holes from the composite to give the completed solid model – Fig. 9.6.
19. SOLLIST the composite.

20. SOLMESH, HIDE, SHADE as previous exercises and then return the model to wire-frame representation with REGEN and SOLWIRE.
21. Make the top left viewport active and restore the UCS FRONT.
22. This viewport has handle 8 (this is my handle – you should have a note of your viewport handle numbers).
23. Set layer DIM-8 current.
24. Dimension this view.
25. Complete the dimension process in the other two viewports, remembering:
    (a) to set the correct UCS for that viewport
    (b) make the correct layer current – handle?
26. Enter paper space, modify your sheet layout and freeze the VPBORDER layer.
27. Your final drawing should be similar to Fig. 9.7 and is now ready for plotting. Figure 9.7 has been plotted with the VPBORDER layer frozen.

This model has used the solid primitives in its construction. Our next exercise will create a solid using the swept primitive SOLEXT.

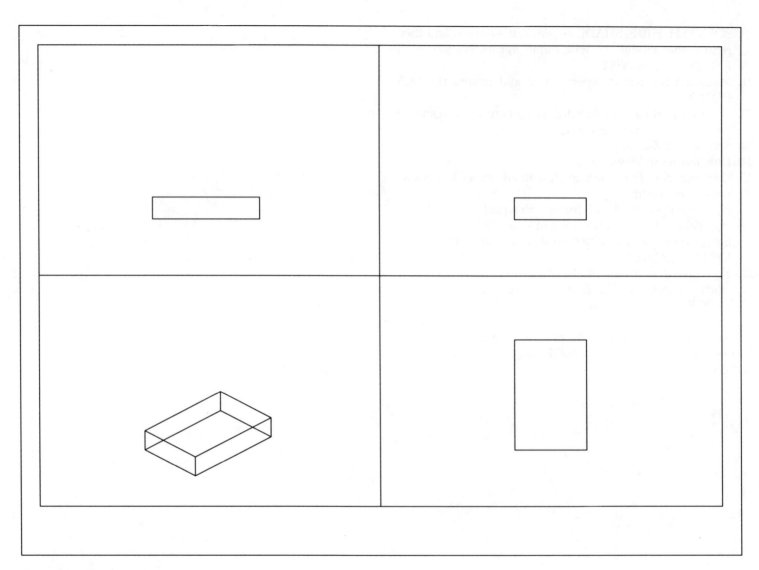

Fig. 9.1. Machine support – base.

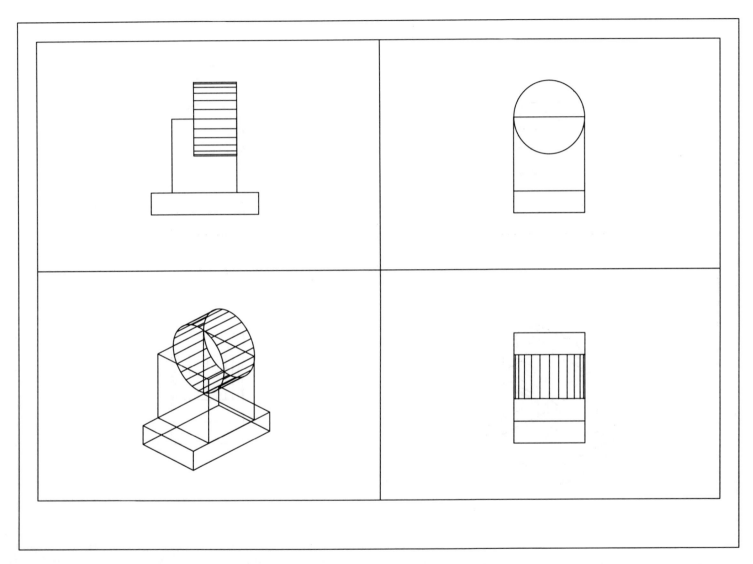

Fig. 9.2. Machine support – after box and cylinder added.

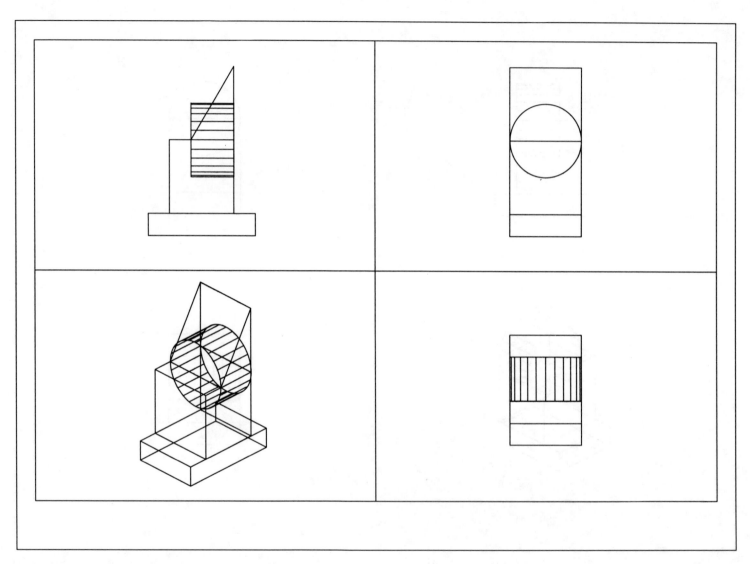

Fig. 9.3. Machine support – after WEDGE added and rotated.

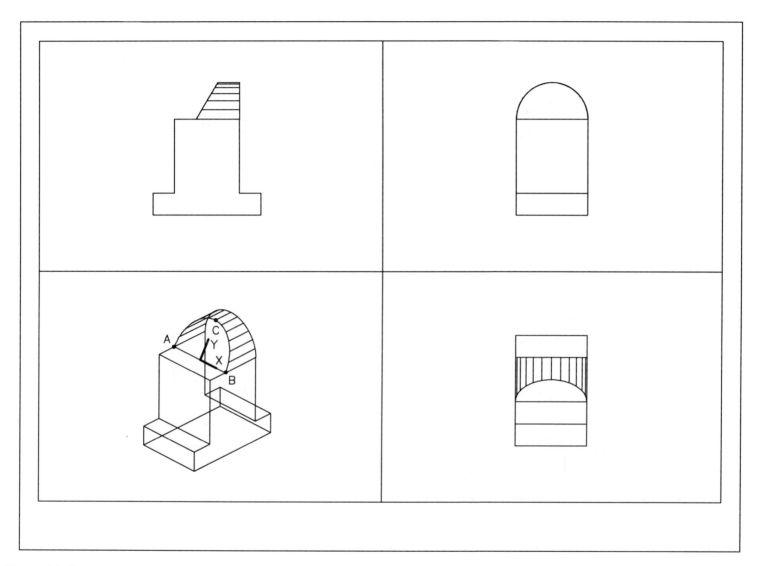

Fig. 9.4. Machine support – UCS slope setting.

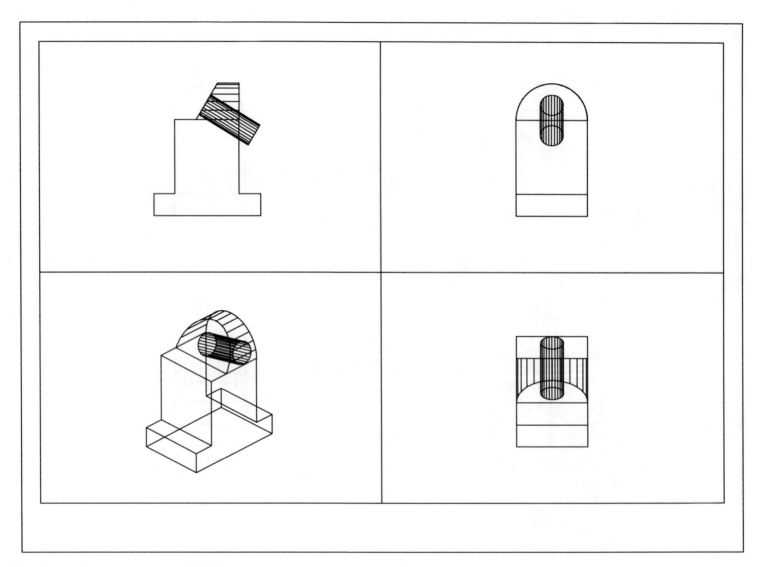

Fig. 9.5. Machine support – after SOLCYL on slope.

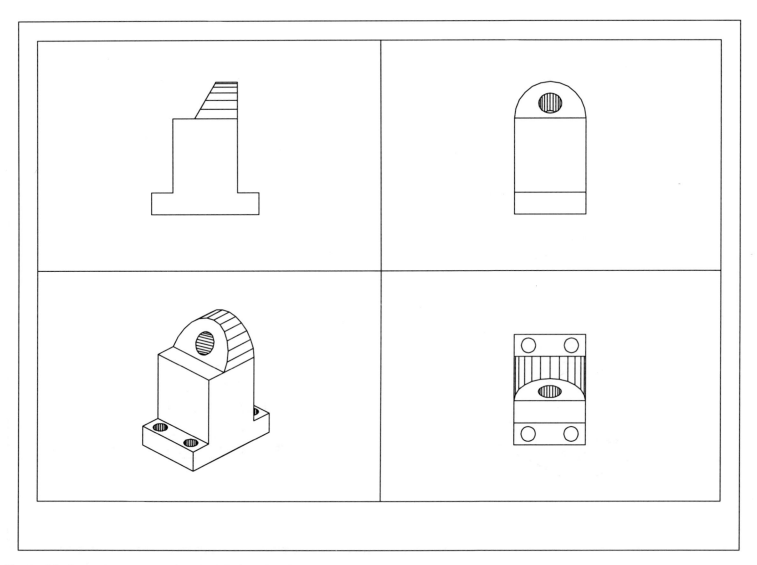

Fig. 9.6. Machine support – completed solid plotted with HIDE.

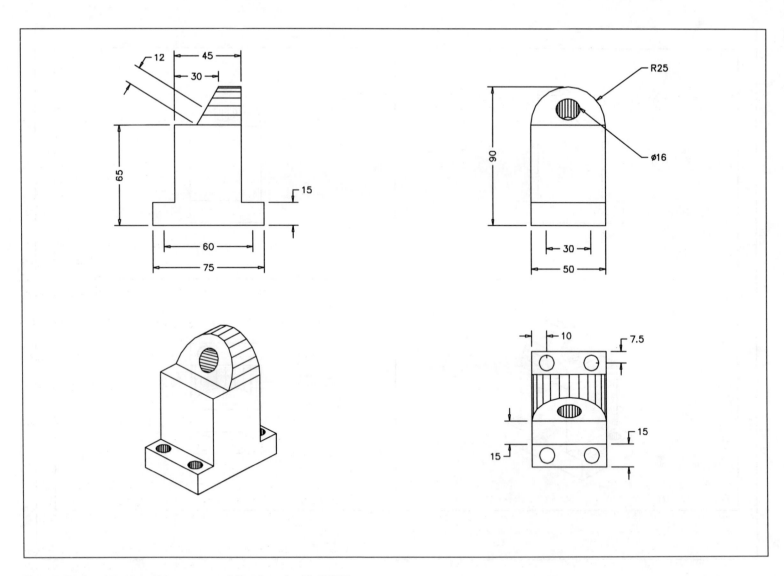

Fig. 9.7. Dimensioned machine support fully plotted with HIDE.

10. Exercise 7: a backing plate

This exercise will use the swept primitive SOLEXT to create the solid model. It will also involve alteration to the SOLA3 standard sheet by changing the viewports. This in itself is a good exercise for the reader, as we do not want to spend time creating another standard sheet.

The model sizes

Refer to Fig. 10.1 which details the model we will create, and gives all relevant sizes. As an aside, could you draw the three views as given in orthographic projection? What about the isometric?

Changing the standard sheet
1. Begin a new drawing **\SOLID\BACKPL=\SOLID\SOLA3**.
2. In paper space use the **STRETCH** command with:

first corner prompt	enter 5,125**<R>**
other corner prompt	enter 375,160**<R>**
base point	pick INT of the 4 viewports at centre of screen
second point	enter @0,–70**<R>**.

3. You should now have two large viewports at the top of your paper and two smaller ones at the bottom.
4. Now ERASE the lower left viewport, and MOVE the left hand viewport to the lower left corner of the small viewport, to give a layout as Fig. 10.2.
5. SAVE this layout as **\SOLID\SOLA3MOD** – we may use it later.

Creating the model
6. In model space, set view points as Fig. 10.2 and ZOOM C about the point **0,10,60** at **1XP** in the three viewports.
7. With the lower right viewport active, use the PLINE command from 0,0 to draw the right-hand half of the backplate outline.

   The sizes should be taken from Fig. 10.1 and you will need to remember to add arc segments as you draw the polyline shape – this outline is more tricky than you would imagine, especially working out the co-ordinates. I used absolute entry!
8. When the half shape is complete, MIRROR it about a vertical line through (0,0) – Fig. 10.2.
9. Use PEDIT to turn the complete outline into a single polyline entity.
10. Change the solid wire mesh density to 3 with **SOLWDENS <R>** and enter **3<R>** at the prompt.
11. Now enter **SOLEXT<R>** and extrude the polyline outline a height of 120 with 0 taper – Fig. 10.3.
12. Restore the UCS LEFT.
13. Make the left viewport active.
14. Draw a construction line from the MIDPOINT of the bottom front edge of the solid, the length being @0,10.
15. Use the PLINE command to draw the 50×10 rectangular cut-out.
16. Erase the red construction line and SOLEXT the rectangle by –6 at 0 taper. The –6 is the plate thickness, negative due to the UCS position.

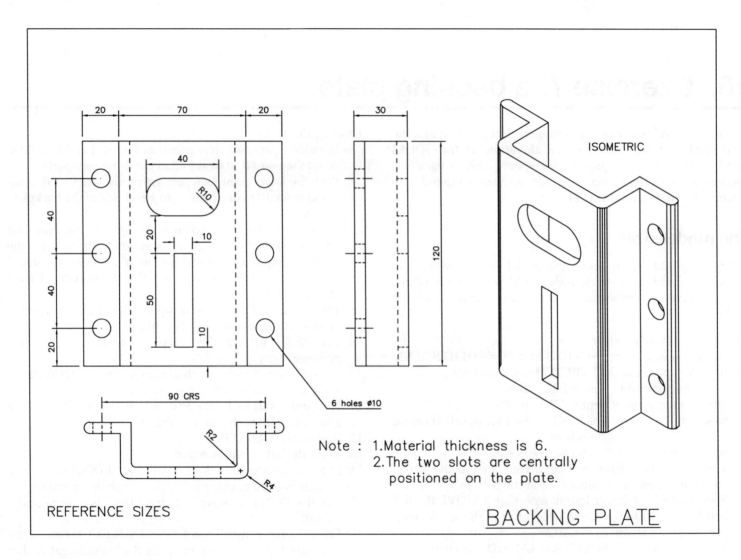

Note : 1. Material thickness is 6.
2. The two slots are centrally
   positioned on the plate.

REFERENCE SIZES

BACKING PLATE

Fig. 10.1. Backing plate – 2D drawing.

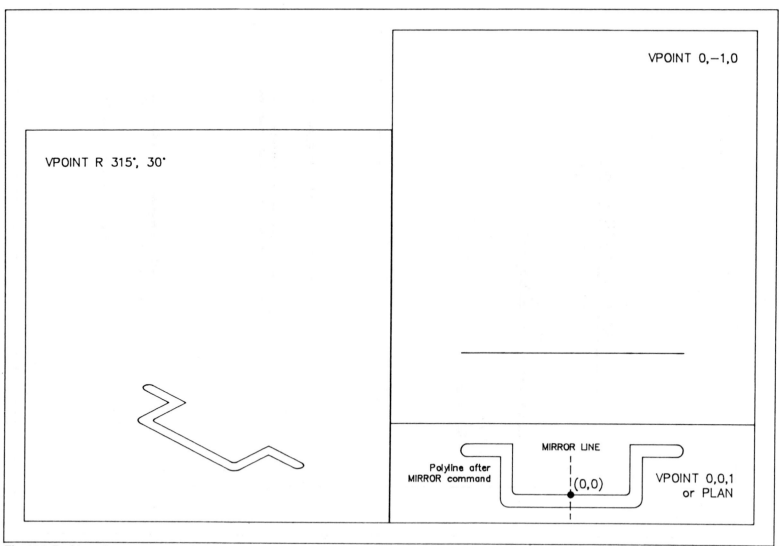

VPOINT 0,−1,0

VPOINT R 315°, 30°

MIRROR LINE

Polyline after
MIRROR command

(0,0)

VPOINT 0,0,1
or PLAN

Fig. 10.2. Backing plate – viewports and polyline shapes.

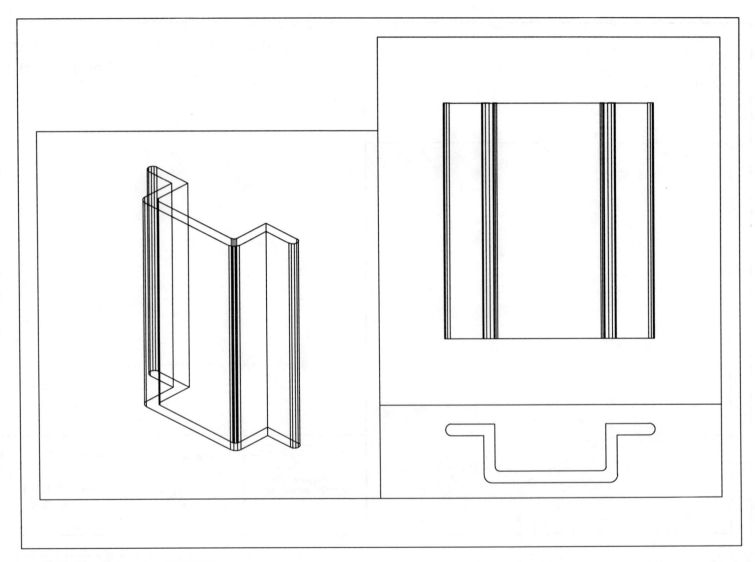

Fig. 10.3. Backing plate – after SOLTEXT.

17. Now use CHPROP and change the colour of this extrusion to green.
18. Alter the SOLWDENS variable to 4.
19. Still in the left viewport, draw a second construction line from the MIDPOINT of the top front face of the rectangle, the length being @0,−20.
20. With PLINE, draw the circular-ended slot – again if you cannot manage this in one operation you will need to PEDIT and join all polylines into one.
21. Erase the construction line and change the colour of the slot outline to blue.
22. SOLEXT the blue slot by −6 with 0 taper – Fig. 10.4.
23. Subtract the two slots from the solid plate.
24. With the UCS still at LEFT, create the first 'hole' with the **SOLCYL** command, using 45,20,−18 as the start point, hole diameter 20 and height −6. Can you reason out the hole centre co-ordinates from the position of the origin in Fig. 10.2. and the sizes in Fig. 10.1?
25. Change the colour of this hole to yellow with CHPROP.
26. Now ARRAY the yellow hole using
   > rectangular
   > 3 rows
   > 2 columns
   > 40 row distance
   > −90 column distance
27. Subtract the six holes from the composite.
28. This completes the solid model – Fig. 10.5.
29. SOLMESH, HIDE and SHADE the composite and the final result is interesting, the coloured cut-outs showing nicely. Return to wire-frame representation with REGEN and SOLWIRE.
30. In paper space, modify your sheet then SAVE. Plot if required.

Investigating the composite

Before leaving this exercise, we will investigate two of the commands which are available to us dealing specifically with solid models. It will also check on the accuracy of your drawing compared to mine.

1. From the screen menu, select **MODEL**
   > **INQUIRY**
   > **SOLAREA**

AutoCAD prompts	`Select objects`
respond	**pick the composite<R>**
AutoCAD prompts	`Surface area of solid is` `39090.28 sq.cm.`

How did you get on? Anything near my figure?
2. Now select **INQUIRY**
   > **SOLMASSP** and pick the composite as before.

AutoCAD prompts	`1 solid selected` `Calculating mass properties`
then	`Mass` **828668.1 gm** `Volume` **105428.5 cu.cm.**

   and much more.
3. We will deal with the mass properties in greater detail in a later chapter, but you will see from the listing that there appears to be a large amount of engineering detail given.
4. At this stage, answer **N<R>** to the plot file prompt, and proceed to the next chapter.

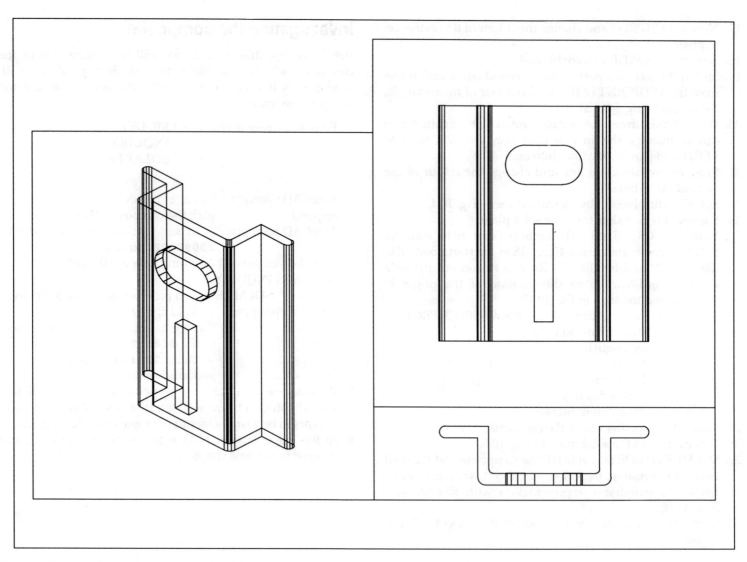

Fig. 10.4. Backing plate – after slots added and SOLTEXT.

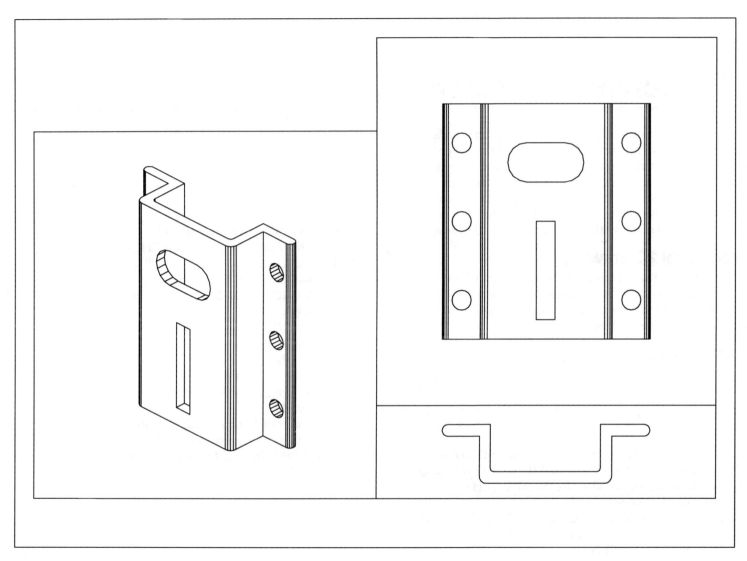

Fig. 10.5. Backing plate – completed composite plotted with HIDE.

11. Exercise 8: a sort of level

This exercise reinforces the solid modelling commands which have been used in previous exercises, but the actual composite is more complicated. To compare solid modelling with traditional 2D drawing is difficult as they are used for entirely different purposes, but it is my intention to show that by the end of the exercise the views obtained from the solid model are easier and more accurate than the comparable 2D views.

Traditional 2D drawing

Refer to Fig. 11.1 which gives the reference sizes for the model to be created. As a diversion, draw the given three views and note how long a took to complete the task. Could you draw an isometric of the component? It is difficult.

Solid model

In this exercise all the relevant data to complete the model will be given, although the prompts will be omitted. You should be familiar with these by now anyway. Don't just accept my figures. Check the input from Fig. 11.1 to see why the values given are being used.

1. Begin a new drawing **\SOLID\LEVEL=\SOLID\SOLA3**.
2. In paper space, MOVE the middle vertical yellow line of the viewports by @–50,0. Use the STRETCH command for this and window this middle line.
3. In model space ZOOM C about 100,9,30 at 0.75XP in the 3D port, and 1XP in the other three viewports.
4. Make the lower right viewport active.
5. Use PLINE to draw the level outline from
   9,–9
   to @182,0
   arc to @0,18
   line to @–182,0
   arc to @0,–18<**R**>
6. SOLEXT this polyline by 60 with 0 taper.
7. Make the extrusion green with CHPROP.
8. To produce the step effect, we will draw boxes and wedges and then subtract these from the solid, so:
   (a) SOLBOX at 0,–20,45 with length 135, width 40 and height 15
   (b) SOLWEDGE at 135,–20,60 and length 15, width 40, height –15
9. Change the colour of this box and wedge to green, then union the two primitives.
10. COPY this box/wedge composite from 150,–20,60 by @–85,0,–15.
11. Now subtract the box/wedge composites from the original solid to give the main body of the level.
12. Still in the lower right viewport, zoom in on the 'centre' of the solid.

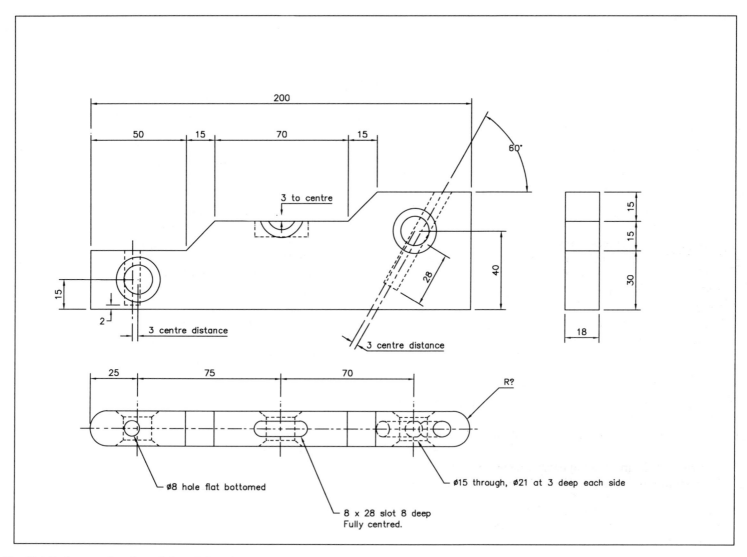

Fig. 11.1. Reference sizes for solid model Exercise 8.

13. Create the top 'hole' with a PLINE
      from 100,–9,45
      to @10.5,0
      to @–3,3
      to @0,12
      to @3,3
      to @–10.5,0<**R**>
14. Revolve this polyline with SOLREV about its two end-points for a full circle.
15. The 'hole' has to be repositioned, so MOVE it with a base point of 100,–9,45 for @0,0,3, i.e. 3 in the Z direction.
16. Multiple COPY this 'hole' from 100,–9,45 to:
 (a) @–75,0,–33 – the hole at the left
 (b) @70,0,–8 – the hole at the right.
17. Subtract the three holes from the composite.
18. SAVE at this stage – you don't want to loose all your work.
19. In the lower left viewport, SOLMESH, HIDE and SHADE then return to wire-frame representation with REGEN and SOLWIRE.
20. In the lower right viewport, create a cylinder at 22,0,30 with diameter 8 and height –28.
21. Create another cylinder at 170,0,70 with a diameter of 8 and a height of –65.
22. In the top right viewport, restore the UCS LEFT.
23. Rotate the last cylinder about the centre of the right hole by –30.
24. Use CHPROP to change the colour of these two cylinder to blue.
25. Return to WCS and the lower right viewport.
26. Subtract the two blue cylinders form the composite.

27. For the slotted hole, use PLINE
      from 90,4,45
      to @20,0
      arc to @0,–8
      line to @–20,0
      arc to close
28. Change this slot to yellow then extrude it by –8 with 0 taper.
29. Subtract the extrusion from the composite and zoom all – this completes the creation of the solid – Fig. 11.2.
30. Using the appropriate viewport, set the relevant DIM-? layer as current and restore the necessary UCS, e.g. LEFT, FRONT? Add the dimensions given in Fig. 11.2 – this should not be too difficult for you.
31. Check your composite values with:
 (a) MODEL, INQUIRY, SOLAREA:
   surface area is 28,497.92 cm^2.
 (b) MODEL, INQUIRY, SOLMASSP:
   mass = 1,057,920 g
   volume = 134,595.4 cm^3.
32. This completes the exercise, so paper space and add text as required to your sheet layout. Save and plot.

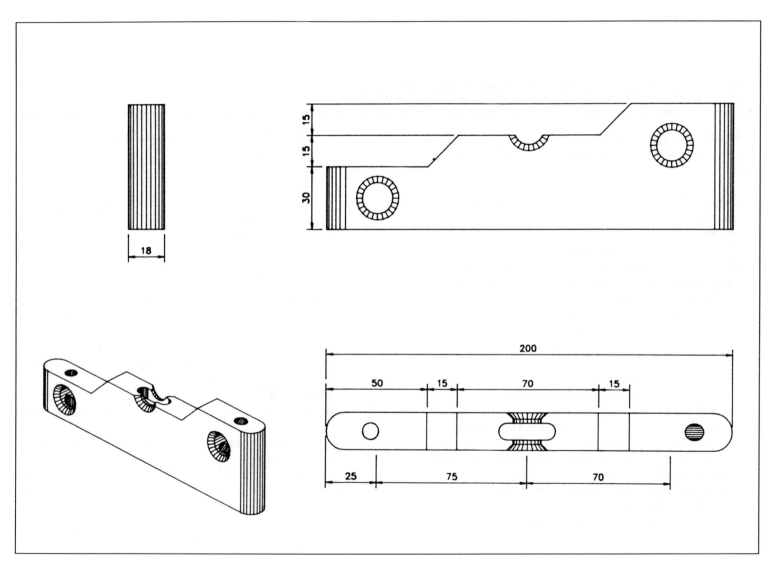

Fig. 11.2. Completed 'sort of level' plotted with HIDE.

12. Exercise 9: pipe and flange

This exercise involves the creation of a pipe and flange joint and should be interesting to the reader. It involves the SOLREV swept primitive.

1. Begin a new drawing **\SOLID\PIPEFL=\SOLID\SOLA3**.
2. In model space ZOOM C about 100,30 at 0.4XP in each viewport.
3. With the bottom left viewport active, restore UCS FRONT.
4. Draw two concentric circles at 0,0 with radii of 30 and 40, respectively. Change colours to make the smaller circle green and the larger circle blue.
5. Draw construction lines from 0,0
    to @200,0
    to @0,100

6. Use the SOLREV command and:
    (a) pick the green circle
    (b) enter E – for entity
    (c) pick the bottom of the vertical red construction line
    (d) enter and angle of revolution of 70°.
7. Repeat the SOLREV command on the blue circle, using the same responses as step 6.
8. Subtract the green pipe from the blue pipe.
9. With the bottom right viewport active, restore WCS and erase the two red construction lines.

10. Use the PLINE command to draw the flange
    from 0,–30
    to @0,–10
    to @–10,0
    arc to @–10,–10
    line to @0,–50
    to –30,0
    to @0,70
    to close.

11. Change the colour of this polyline to yellow.
12. SOLREV the yellow polyline about the points 0,0 and –50,0 for a full circle.
13. In the top left viewport, restore UCS FRONT.
14. Create a cylinder using SOLCYL at the point 0,75,–20 with a diameter of 20 and a height of –30.
15. CHPROP and alter the colour of this cylinder to magenta.
16. Polar array the cylinder for 6 items about the point 0,0 for a full circle.
17. Subtract the six holes from the flange, and then union the flange and the pipe.
18. This completes the drawing (Fig. 12.1). In the lower left viewport use SOLMESH, HIDE and SHADE to give a very spectacular effect. Return the model to wire-frame representation with REGEN and SOLWIRE.
19. Enter SOLLIST and pick the composite object. The screen will prompt with UNION and give the material as MILD-STEEL.

20. Enter SOLMASSP and pick the composite. The screen displays all the calculated mass properties with: mass = 10,321,827 g and volume = 1,313,210 cm^3.
21. We will now investigate how the material can be changed, so select form the screen menu **MODEL**

UTILITY

DDSOLMAT

   (a) the screen displays the Materials dialogue box.
   (b) pick BRASS from the Materials column.
   (c) pick Load>
   (d) pick BRASS from the Material in Drawing column
   (e) pick Set
   (f) pick Change and select the composite**<R>**
   (g) pick the OK box.
22. Now enter SOLLIST and select the composite – UNION, BRASS.
23. Enter SOLMASSP again and select the composite. The properties for comparison are: mass = 11,122,885 g and volume = 1,313,210 cm^3.

   We will investigate the mass properties in detail in a later chapter.
24. That completes this exercise, so in paper space modify your sheet layout, save and plot.

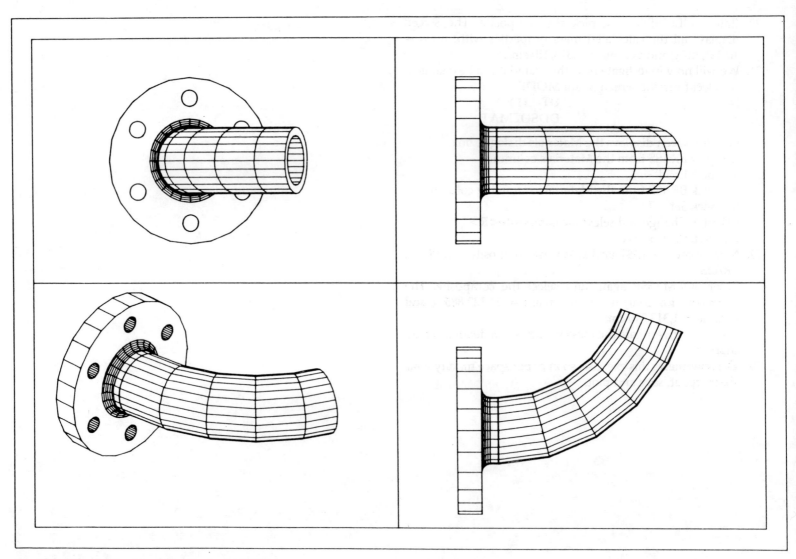

Fig. 12.1. Pipe and flange solid plotted with HIDE.

13. Review

We have now completed nine different solid modelling exercises, and I hope that you are now beginning to get to grips with the commands used. There have been ten types of primitives used in the exercises and these are usually grouped as follows:

Classic geometry	*Swept*	*Edge*
Box	Extrusion	Chamfer
Wedge	Revolution	Fillet
Cylinder		
Cone		
Sphere		
Torus		

Classic geometry primitives

These are the AutoCAD 'building blocks' for solid modelling. They allow the user to create a variety of shapes by contrasting user entry

(a) a box can be either a cube, a thin plate, a long narrow box or a regular type of cuboid
(b) the wedge shape can be regular, a right triangle, long tapered or long and non-tapering
(c) cylinders can be regular, disk or rod shaped, as well as being elliptical
(d) cones can be made regular, flat or needle pointed
(e) the sphere is limited to a sphere of varying sizes
(f) the torus can be large or small with differing tube radii.

Swept primitives

These allow greater flexibility than the classic primitives. Complex shapes can be drawn in 2D and extruded or revolved to duplicate virtually any 3D shape – consider the splined shaft, moulding and pulley examples.

Edge primitives

These are limited and never used as 'stand alone primitives' but as addition to existing solids. They are convenient rather than flexible.

Irregular surfaces

Solids which have irregular surfaces to do mesh easily using constructive solid geometry, and this is perhaps the only disadvantage of using AutoCAD for solid modelling. The problem occurs due to the inability of AutoCAD to create solids from surfaces with irregular solids. For this reason, solids with irregular surfaces are better constructed using boundary representation.

Thin shells

This is not well supported with AutoCAD. Although thin shell solids can be created, the SOLWDENS variable may not work properly.

Efficient modelling

There is no one way to produce a solid model. The number of primitives should be kept as low as possible, and swept primitives are better than classic primitives.

14. Regions

A region is a 2D area defined by an outer perimeter, but to the user it looks very much like the cross-section of a 3D object. A region has the following characteristics:

- it is a solid of zero thickness in the Z direction
- can contain inner perimeters called **loops**, e.g. holes
- is confined to a single plane
- the loops (outer and inner) are continuous
- every region has one outer loop
- there may be one or more inner loops – **composite regions**
- inner loops are in the same plane as the outer loop
- inner loops never intersect
- an inner loop is contained entirely within an outer loop
- they can be extruded (SOLEXT) or revolved (SOLREV)
- loops can be extruded to different heights/depths
- loops can be revolved to different angles, positive/negative
- perimeters are drawn using polylines, circles, etc
- regions are 'solidified' with the SOLSOLIDIFY command, or become solid automatically after SOLEXT or SOLREV
- regions and solids can not be mixed
- regions must lie on the same plane for the union operation
- regions have a CSG tree.

Using regions does not allow the user any new ability to create solid models. It is another variation to be considered. In fact we have used regions in our previous exercises without knowing about them, i.e. every time a polyline shape has been extruded or revolved, the original polyline shape has been a region. The extrusion, revolution command turned the region into a solid.

Region examples

Regions will be demonstrated by two worked examples and in the process change the appearance of our standard sheet. We will also introduce the solid hatch commands.

1. Begin a new drawing **\SOLID\REGEX1**=**\SOLID\SOLA3**.
2. In paper space with the layer VPBORDER current, erase the four viewports.
3. Enter MVIEW, and create two vertical viewports with: first point at 10,25 second point at 370,260
4. In model space, make layer SOLIDS current, and set viewpoints as follows:
    left VPOINT 0,0,1
    right VPOINT R 315 30
5. ZOOM C about 0,0 at 1XP.
6. Enter the following solid hatch values:
    SOLHPAT: U
    SOLHSIZE: 3
    SOLHANGLE: 45
7. Save this drawing as **\SOLID\REGSTD**.

Example 1

1. Refer to Fig. 14.1 and draw the letter 'R' from three different polylines. The sizes are unimportant – use your discretion.
2. Subtract the two smaller shapes from the large R using the SOLSUB command. When complete, AutoCAD will prompt:

   ```
   1 region selected
   2 regions selected
   2 regions subtracted from 1 region
   ```

3. The region will be hatched as Fig. 14.1.
4. Enter SOLEXT and pick the region. Enter an extruded height of 30 and a taper angle of 0. Enter **N** to the extrude loops prompt and Fig. 14.2 should result.
5. Undo the SOLEXT command with **U<R>**.
6. Zoom in on the 'top' of the R.
7. Repeat the SOLEXT command, pick the region and enter 30 and 0 as before.

AutoCAD prompts	`Extrude loops to different heights`
enter	**Y<R>**
AutoCAD prompts	`pick loops for new height`
respond	pick the outer loop of the R then**<R>**
AutoCAD prompts	`Height of extrusion`
enter	**–10<R>**
AutoCAD prompts	`pick loops for new height`
respond	pick the inner 'D' loop then**<R>**
AutoCAD prompts	`Height of extrusion`
enter	**50<R>**
AutoCAD prompts	`pick loops for new height`
enter	**<RETURN>**

8. The result should be as Fig. 14.3 the extrusion having the 'D' of the letter R extruded above the rest of the letter.
9. Save your drawing.

Example 2

1. Begin a new drawing **\SOLID\REGEX2=\SOLID\REGSTD**.
2. Refer to Fig. 14.4 and draw a circle at 0,0. Draw other circles and array then for 4 items and 12 items as shown.
3. Subtract the smaller circles from the larger circle to create the region – Fig. 14.4.
4. Extrude the region for a height of 50, with 0 taper and N to the loops prompt – Fig. 14.5.
5. Undo the extrusion then repeat it on the region, but answer Y to the loops prompt. I extruded to outer loop by –10, and the four inner loops (circles) by 10, 20, 30 and 40, respectively.
6. The result was Fig. 14.6, which I viewed at VPOINT R 200 20 for effect.

This completes our introduction to regions. Remember regions do not have to be used to create solids. They offer you another alternative, but if you can create your solids from the primitives, then regions need not be used.

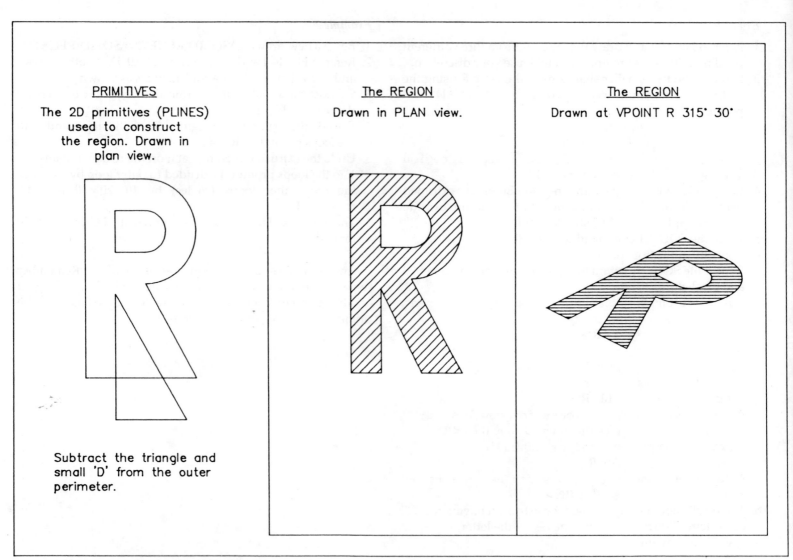

PRIMITIVES

The 2D primitives (PLINES) used to construct the region. Drawn in plan view.

Subtract the triangle and small 'D' from the outer perimeter.

The REGION

Drawn in PLAN view.

The REGION

Drawn at VPOINT R 315° 30°

Fig. 14.1. Region example 1 – after the SOLSUB command.

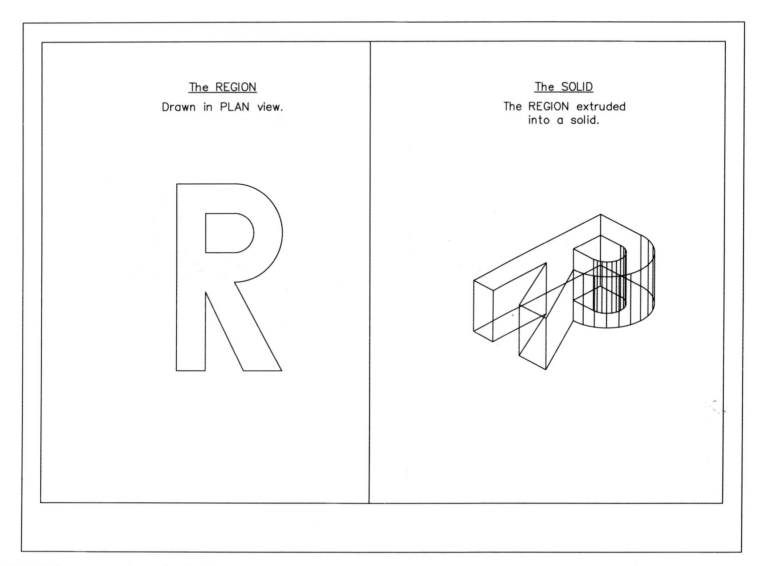

The REGION

Drawn in PLAN view.

The SOLID

The REGION extruded
into a solid.

Fig. 14.2. Region example 1 – after the SOLTEXT command.

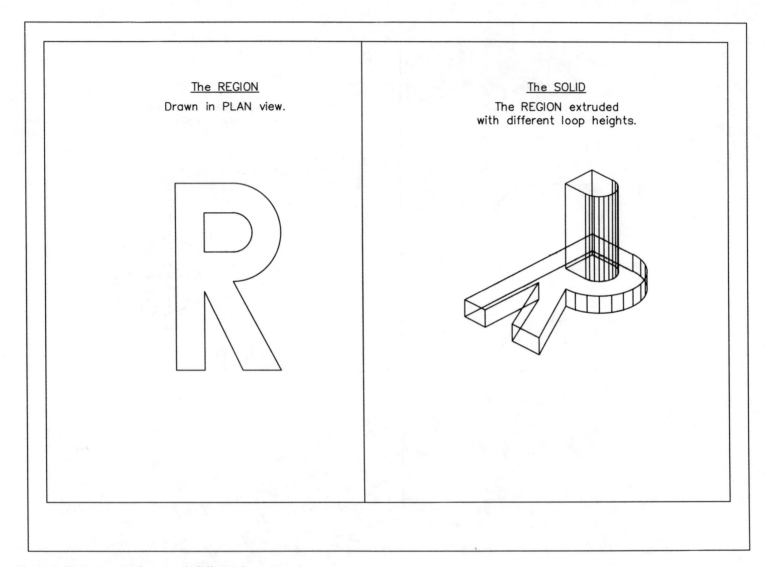

The REGION
Drawn in PLAN view.

The SOLID
The REGION extruded
with different loop heights.

Fig. 14.3. Region example 1 – with different loop extrusions.

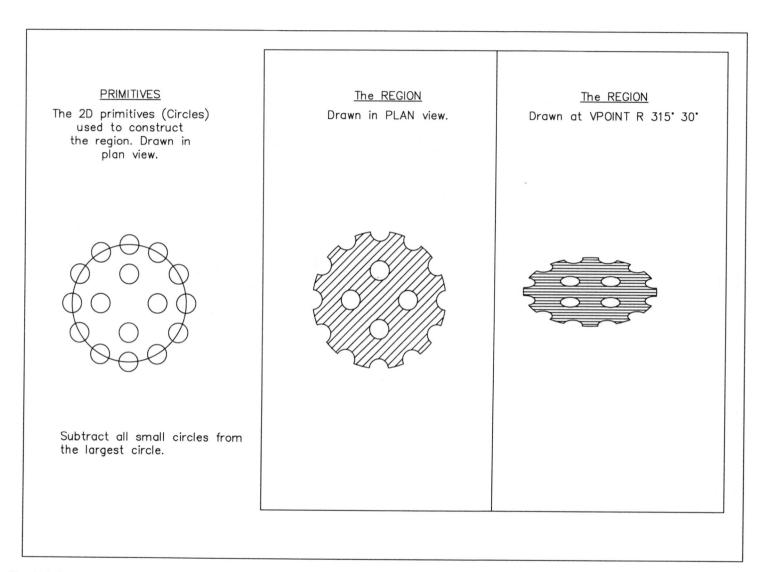

Fig. 14.4. Region example 2 – the region after SOLSUB.

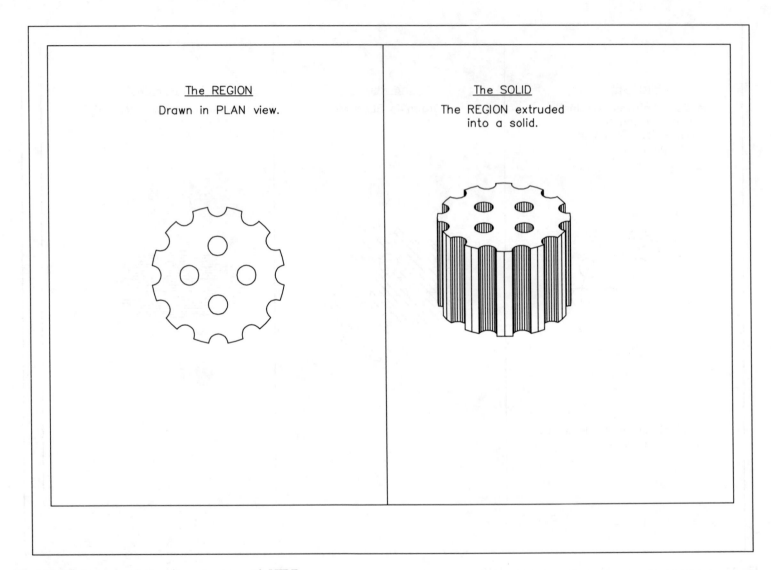

The REGION
Drawn in PLAN view.

The SOLID
The REGION extruded
into a solid.

Fig. 14.5. Region example 2 – after extrusion with HIDE.

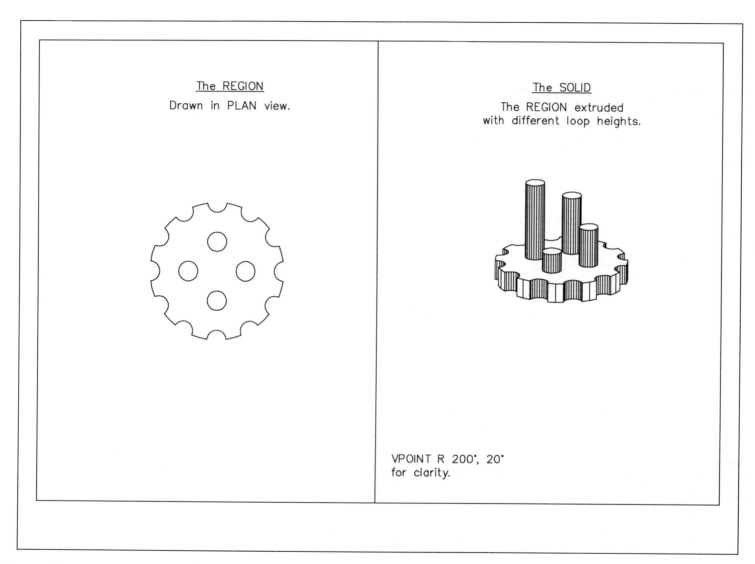

The REGION
Drawn in PLAN view.

The SOLID
The REGION extruded
with different loop heights.

VPOINT R 200°, 20°
for clarity.

Fig. 14.6. Region example 2 – after different extrusions plotted with HIDE.

15. Moving solids

Solids can be moved and rotated with the traditional MOVE and ROTATE commands, although the actual result may not be as expected due to the UCS position. To facilitate moving and rotating solid objects, AutoCAD allows the user access to the SOLMOVE command.

SOLMOVE – will allow the user to perform several different operations:

1. Translate or rotate a solid.
2. Reposition the MCS icon.
3. Reposition both the icon and the solid.

One advantage of the SOLMOVE command over the MOVE command is that a solid can be rotated without having to set up an appropriate UCS – think about the ROTATE command!

The motion co-ordinate system (MCS)

When the SOLMOVE command is used and the object selected, a special temporary icon appears on the screen in all viewports and at the origin of the current UCS. This icon is for the motion co-ordinate system (MCS), and is different from either the WCS or UCS icons. It has a single arrow on the X axis, a double arrow on the Y axis, and three arrows on the Z axis. These arrows show the orientation of the MCS's X, Y and Z axes. When the icon appears on the screen, the user is repeatedly prompted to enter '**Motion descriptions**' allowing a variety of options, e.g. move the icon, move the solid, etc.

We will demonstrate the SOLMOVE by example, then discuss in detail the options which are available with the command.

SOLMOVE example 1

This exercise will move one solid onto another by the traditional MOVE and ROTATE commands as well as the SOLMOVE command.

1. Begin a new drawing **SOLID\MOVEX1**=**SOLID\SOLA3** and refer to Fig. 15.1.
2. In paper space, set the VPBORDER as current and ERASE the four viewports.
3. With MVIEW create a single viewport from 10,25 to 370,260.
4. In model space, set VPOINT R 315 30.
5. Create the following primitives:
   (a) a wedge at 20,20,0 with length 50, width 35 and height 25
   (b) a cylinder at 120,70,0 with diameter 35 and height 15.
6. Multiple COPY the two primitives

	from	20,20,0	– diagram (A)
	to	@0,0,–100	– diagram (B)
	to	@0,0,–200	– diagram (C)
	to	@0,0,–300	– diagram (D)

7. Now ZOOM in on the eight items.

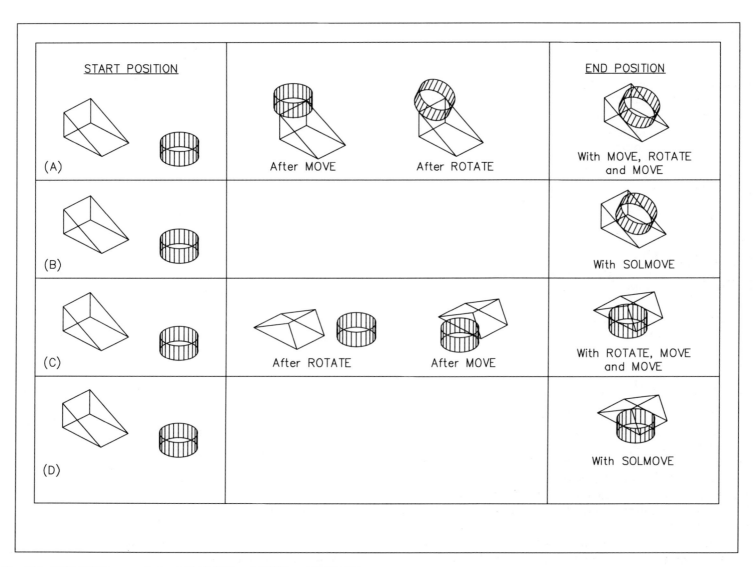

Fig. 15.1. SOLMOVE example 1 – SOLMOVE vs MOVE and ROTATE.

A. Cylinder on slope with MOVE and ROTATE

1. MOVE the cylinder from the centre of its base, to the mid-point of the top edge of the wedge – OSNAP obviously.
2. Using 3 point, set the UCS to the corner of the wedge side.
3. ROTATE the cylinder about the top edge of the wedge, so that the base is 'flush' with the slope of the wedge – use OSNAP END and pick the bottom corner of the slope at the rotation angle prompt
4. Restore the WCS, and MOVE the cylinder half way down the slope.

B. Cylinder on slope with SOLMOVE

At the command prompt, enter **SOLMOVE<R>**

AutoCAD prompts	Select object
respond	**pick the cylinder**
AutoCAD prompts	1 solid selected
	?/<Motion description>
enter	**E<R>** – for edge
AutoCAD prompts	Select edge to define co-ordinate system
respond	**pick the base circle of the cylinder**
AutoCAD prompts	with a green(?) MCS icon, origin at the base circle of the cylinder and aligned with the WCS icon
then	?/<Motion description>
enter	**AE<R>** – align icon and cylinder
AutoCAD prompts	Select edge to define co-ordinate system
respond	**pick near slope edge of wedge**
AutoCAD prompts	by moving the icon
then	No/<Yes>
enter	**Y<R>**
AutoCAD prompts	by moving the cylinder to the edge (wrong way?)

then	?/<Motion description>
enter	**RX-90<R>**
then enter	**TX-17.5<R>**
then enter	**<RETURN>** to end sequence.

The cylinder should have been moved to the centre of the wedge slope and be lying along the slope.

Note

1. After the AE entry and the edge selection, AutoCAD may give a prompt – REPEAT SELECTION. All you need to do is pick the edge again.
2. The RX-90 and TX-17.5 entries can be entered in one line as **RX-90,TX-17.5<R>**.

C. Wedge on cylinder with MOVE and ROTATE

1. Create a vertical UCS about the three endpoints of the near triangular side of the wedge.
2. ROTATE the wedge by 207° about 0,0 – aligns the slope surface of the wedge with the 'horizontal' plane.
3. Restore the WCS.
4. MOVE the wedge from the midpoint of one flat edge to the centre of the top surface of the cylinder.
5. Centre the wedge on the cylinder with MOVE command and enter @0,–17.5,0 ?

D. Wedge on cylinder with SOLMOVE

At the command prompt, enter **SOLMOVE<R>**

AutoCAD prompts	Select objects
respond	**pick the wedge**
AutoCAD prompts	1 solid selected
then	?/<Motion description>
enter	**F<R>** – face option
AutoCAD prompts	Select face to define co-ordinate system

respond	pick the near edge of the slope of the wedge
AutoCAD prompts	`Next/<OK>`
respond	**Enter N (if needed) until the MCS icon is aligned with the slope then <RETURN>**
AutoCAD prompts	`?/<Motion description>`
enter	**AE<R>** – align icon and wedge
AutoCAD prompts	`Select edge to define destination coord system`
respond	**pick the top edge of the cylinder**
AutoCAD prompts	`by moving the MCS icon to top circle centre then: No/<Yes>`
enter	**Y<R>**
AutoCAD prompts	`by moving the wedge which should not be aligned`
then	`?/<Motion description>`
enter	**RX180<R>**
then enter	**RZ90<R>**
then enter	**TX-17.5<R>**
then enter	**TY-25<R>**
then enter	**<RETURN>** to end sequence

Note

1. The 'repeat selection' prompt may occur again after the AE entry – just pick the top circle of the cylinder again.
2. The motion description codes can be enter as one line, i.e. **RX180,RZ90,TX-17.5,TY-25<R>**.

Motion description codes

The worked example just completed used some of the motion description codes which are available. Letters are used for the motion requirements, and these can be summarised by entering **?** at the 'motion description' command prompt. The complete 'set' of codes can be grouped according to their use as follows:

Options	*Motion description*
1. Translation of solid along the MCS axes	
TX	translation along the X axis
TY	translation along the Y axis
TZ	translation along the Z axis
2. Rotation of solid about the MCS axes	
RX	rotation about the X axis
RY	rotation about the Y axis
RZ	rotation about the Z axis
3. Reorientation of the MCS icon	
E	set the MCS icon to an edge
F	set the MCS icon to a face
U	set the MCS icon to the current UCS
W	set the MCS icon to the WCS
4. Align a solid and the MCS icon	
AE	align with an edge
AF	align with a face
AU	align with the current UCS
AW	align with the WCS
5. Restore the original 'set up'	
O	restore MCS icon and solid to original position

Response descriptions

There are four types of entry available to the user in response to the motion description prompt:

1. **T** or **R** to move (translate or rotate) only the solid in relation to the MCS's current position.
2. **E,F,U** or **W** to move only the MCS icon in relation to any solid, or in relation to the current UCS or WCS.
3. **AE,AF,AU** or **AW** to align both the MCS icon and the object with any solid, or the current UCS or WCS.
4. **O** to restore the object and the MCS icon to the positions they were in prior to the SOLMOVE command.

Moving only the object

(a) To translate an object, enter
> 1. T
> 2. the axis, i.e. X, Y or Z
> 3. the distance (+ or −)

e.g. (i) TX50 to move an object 50 units in the X direction.
> (ii) TY-30,TZ20 to move 30 units in the negative Y direction then 20 units in the positive Z direction.

(b) To rotate an object, enter
> 1. R
> 2. the axis of rotation, i.e. X, Y or Z
> 3. the angle (+ or −)

e.g. (i) RZ25 to rotate 25° about the Z axis.
> (ii) RX70,RY30 to rotate 70° about the X axis then 30° about the Y axis.

(c) Combined translation and rotation is obtained with TX34, RZ56 for example.

Moving only the MCS icon

The axes icon can be moved relative to any solid, or the current UCS or the WCS using the letter codes E, F, U or W. Only the icon moves, and any objects selected are 'left behind'.

(a) To attach the icon to the midpoint of an edge:
> 1. enter E
> 2. pick the edge

(b) To 'attach' the icon to a face:
> 1. enter F
> 2. pick the face

(c) To orientate the icon at the origin of the current UCS, enter U.

(d) To orientate the icon at the WCS origin, enter W.

Moving both icon and object

To move both icon and object:
> 1. enter A
> 2. followed by E or F or U or W.

SOLMOVE example 2

1. Begin a new drawing \SOLID\MOVEX2=\SOLID\SOLA3 and refer to Fig. 15.2.
2. ZOOM C about 50,50,20 at 0.5XP in all viewports.
3. With the lower left viewport active (layer SOLIDS) create the following primitives:
   (a) SOLBOX at 0,0,0 with length 100, width 100, height 20.
   (b) SOLCYL at 150,150,0 with diameter 50 and height 40.
   (c) SOLWEDGE at 200,200,0 with length 40, width 40 and height 50.
   (d) SOLCONE at 200,0,0 with diameter 40 and height 60.
4. Zoom in on the four solids.
5. Enter SOLMOVE and:
   (a) pick the cylinder**<R>**
   (b) enter **E** and pick the bottom circle of the cylinder**<R>**
   (c) enter **AE** and pick the top surface of the cuboid**<R>**
   (d) enter **Y** then **RX-90,TX-50<R>**.
6. Repeat the SOLMOVE command, and:
   (a) pick the wedge**<R>**
   (b) enter **F** and pick the bottom surface of the wedge**<R>**
   (c) N?? if needed then**<R>**
   (d) enter **AE** and pick the top surface of the cylinder
   (e) enter **Y** then **RX180,TY-20,TX-20<R>**
7. SOLMOVE again and:
   (a) pick the cone**<R>**
   (b) enter **E** and pick the base of the cone**<R>**
   (c) enter **AE** and pick the slope of the wedge**<R>**
   (d) enter **Y** then **RX90,TX-20<R>**
8. Use CHPROP to change the colours of the solids, then union all four solids.
9. SOLMESH, HIDE and SHADE then return to wire-frame with ??

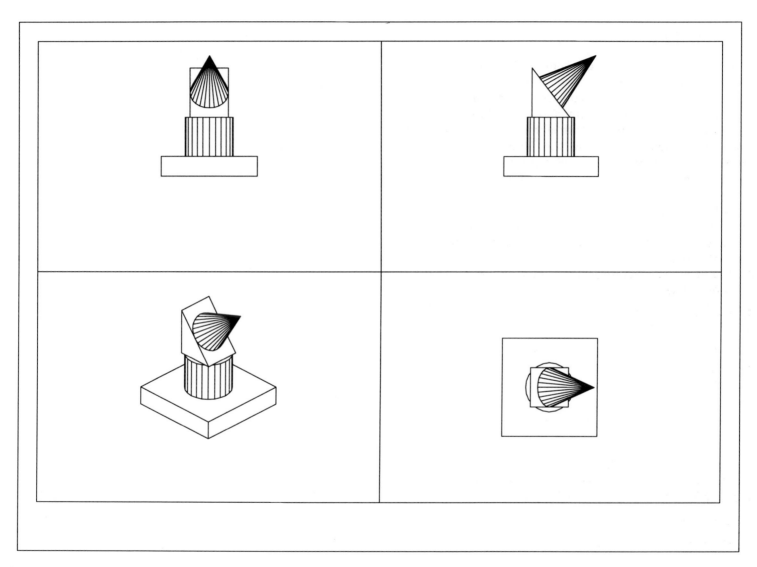

Fig. 15.2. SOLMOVE example 2 using four primitives plotted with HIDE.

❏ *Summary*

The SOLMOVE command does not have to be used to move solid objects as the MOVE and ROTATE commands are quite valid. SOLMOVE gives the user another alternative. It may seem a very 'clumsy command' and it does take some practice to become familiar with all the options available. If you are content with using the MOVE and ROTATE commands, then SOLMOVE will not interest you. When I started solid modelling, SOLMOVE was a command I tried to avoid using as I thought it rather awkward. I would suggest that you created some solids and use the command as example 2. It is a very useful command to have, especially if an object is to be moved 'onto' another object and rotated into a set position.

In our worked examples, the command was selected by entering SOLMOVE at the command line. This is the easiest way of activating the command, but it can be activated from the screen menu and menu bar with:

Screen	Menu bar
MODEL	Model
MODIFY	Modify >
SOLMOVE:	Move object

16. Changing solids

Primitives, regions and composite solids can be moved, copied, rotated, scaled and arrayed with the normal editing commands, but they cannot be stretched. Composites and regions can be changed internally with the **SOLCHP** (SOLid CHange Primitive) command.

SOLCHP example 1

1. Begin a new drawing **???=\SOLID\SOLA3**.
2. In model space ZOOM C about 25,30,50 at 0.6XP in all viewports.
3. With the lower left viewport active, create the following primitives:
   (a) SOLBOX at 0,0,0 with length 50, width 60 and height 80
   (b) SOLCYL at 50,30,40 with diameter 40 and use the C option, entering @50<0<0 as the centre of the other end.
4. Now MOVE the cylinder from 50,30,40 by @−50,0,0.
5. At the prompt line, enter **SOLCHP<R>**.

AutoCAD prompts	Select a solid or region
respond	**pick the box**
AutoCAD prompts	`1 solid selected`
then	Colour/Delete/Evaluate/Instance /Move/Next/Pick/Replace/Size/eXit
enter	**C<R>**
AutoCAD prompts	`New colour <1 (red)>`
enter	**2<R>** – for yellow
AutoCAD prompts	`Colour.................`

enter	**M<R>**
AutoCAD prompts	`Base point or displacement`
enter	**0,0<R>**
AutoCAD prompts	`Second point of displacement`
enter	**@−50,0,0<R>**
AutoCAD prompts	`Colour.................`
enter	**S<R>**
AutoCAD prompts	`with the MCS icon at 0,0,0?`
then	Length along X axis <50>
enter	**10<R>**
AutoCAD prompts	Length along Y axis <60>
enter	**100<R>**
AutoCAD prompts	Length along Z axis <80>
enter	**50<R>**
AutoCAD prompts	`Colour.................`
enter	**X<R>**

6. Repeat the SOLCHP command and pick the cylinder, and at the prompts enter:
   (a) **C** then 3 for green.
   (b) **M** from 50,30,40 to −40,50,25
   (c) **S** then 25 for radius along X axis
            5 for radius along Y axis
            80 for length along Z axis
   (d) **X** to end sequence.
7. Your solid should now be as Fig. 16.1.

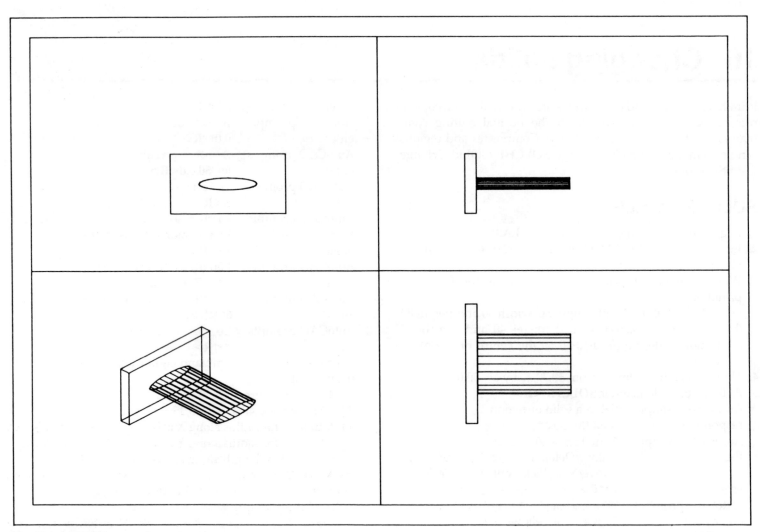

Fig. 16.1. SOLCHP example 1 – two primitives.

SOLCHP example 2

1. Begin a new drawing ???=\SOLID\SOLA3.
2. ZOOM C about 50,50,50 at 0.5XP in all viewports.
3. Create the following:
   (a) SOLBOX at 0,0,0 with the cube option and length 100
   (b) SOLBOX at 0,0,0 with length 20, width 20 and height –30.
4. Rectangular array this last box with two rows and two columns, the row and column distances both being 80.
5. Create a cylinder at 15,50,100 with diameter 15 and height 30.
6. From the menu bar, select **Construct**
   Array 3D
   then
   (a) pick the cylinder**<R>**
   (b) P for polar
   (c) six items
   (d) 360°
   (e) Y for rotation
   (f) 50,50,100 as centre point of array
   (g) 50,50,0 as the second point of array
7. Now union all 11? primitives – Fig. 16.2.
8. From the screen menu, select **MODEL**
   MODIFY
   SOLCHP:

AutoCAD prompts	Select a solid or region
respond	**pick the composite**
AutoCAD prompts	1 solid selected
then	Select primitive
respond	**pick the most left cylinder on the top**
AutoCAD prompts	Colour....................
enter	**C<R>** then **2<R>** for yellow
enter	**S<R>** then 5 for radius along X axis
	5 for radius along Y axis
	50 for length along Z axis

AutoCAD prompts	Colour....................
enter	**N<R>** repeatedly until the OPPOSITE CYLINDER is highlighted
then enter	**C<R>** then **3<R>** for green
enter	**S<R>** then 10 for radius along X axis
	20 for radius along Y axis
	5 for length along Z axis
AutoCAD prompts	Colour....................
enter	**X<R>**

9. Create a cylinder at 150,150,0 with a diameter of 140 and a height of 100.
10. From the menu bar, select **Model**
    Modify >
    ChangePrim

AutoCAD prompts	Select a solid or region
respond	**pick the original composite**
AutoCAD prompts	1 solid selected
then	Select primitive
respond	**pick the large cube**
AutoCAD prompts	Colour....................
enter	**R<R>**
AutoCAD prompts	Select solid to replace primitive
respond	**pick the cylinder created**
AutoCAD prompts	Retain detached primitive
enter	**N<R>**
AutoCAD prompts	Colour....................
enter	**M<R>** and the cylinder is highlighted
AutoCAD prompts	Base point or displacement
enter	**150,150,0<R>**
AutoCAD prompts	Second point of displacement
enter	**50,50,0<R>**
AutoCAD prompts	Colour....................
enter	**N<R>** repeatedly until the 'left-most foot' is highlighted

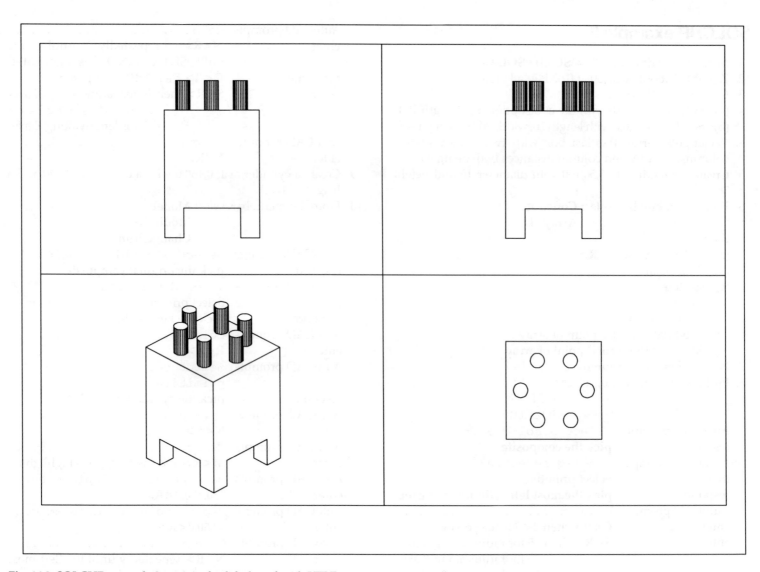

Fig. 16.2. SOLCHP example 2 – original solid plotted with HIDE.

then enter   **M<R>** and move from **0,0** by **@–20,–20**
AutoCAD prompts   Colour.....................
enter   **S<R>** and **50** along the X axis
                               10 along the Y axis
                               –50 along the Z axis
AutoCAD prompts   Colour.....................
enter   **N<R>** until the 'opposite foot' is highlighted
then   **M<R>** and move from **0,0** by **@20,20**
then   **S<R>** and **50** along the X axis
                               50 along the Y axis
                               –5 along the Z axis
then   **X<R>**

11. Your final composite should be as Fig. 16.3.

The complete list of options available with the SOLCHP command is as follows:

Option	Description
(C)olour	Change the colour of a primitive
(D)elete	Delete a primitive from the CSG tree
(E)valuate	Recompose a solid
(I)nstance	Make a copy of a primitive
(M)ove	Move a primitive
(N)ext	Select the next primitive
(P)ick	Pick a primitive
(R)eplace	Replace a primitive with a solid
(S)ize	Change the dimensions of a primitive
e(X)it	Exit the SOLCHP command

I have demonstrate several of these options in the worked examples, and will leave the reader to investigate the rest at their leisure.

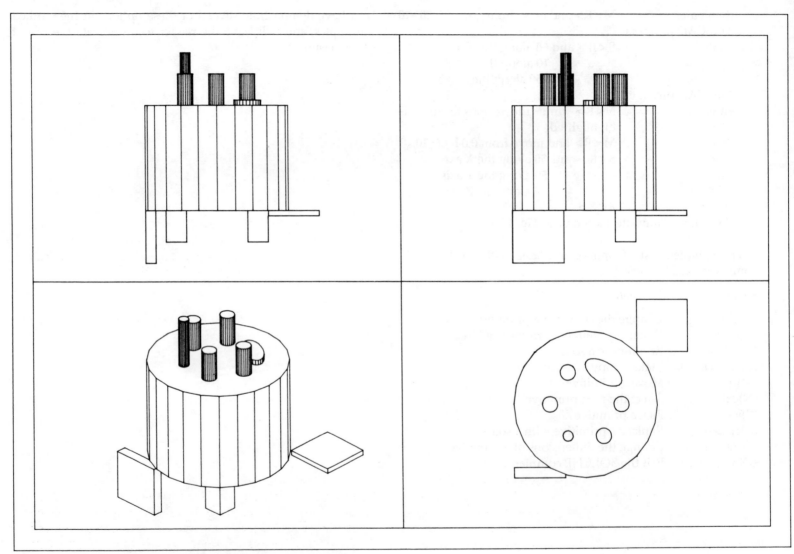

Fig. 16.3. SOLCHP example 2 – after SOLCHP plotted with HIDE.

17. Exercise 10: a casting block

This exercise will use the SOLMOVE and SOLCHP from the previous sections. While all steps are included, the prompts will be shortened, as the reader should be starting to become familiar with them.

The basic casting

1. Begin a new drawing \SOLID\CASTING=\SOLID\SOLA3.
2. Set SOLWDENS to 5.
3. Use SOLBOX to create a cuboid at 0,0,0 with a length of 75, a width of 100 and a height of 36.
4. ZOOM C about 30,50,18 at 1XP in all viewports.
5. Create a cylinder (SOLCYL) at 0,0,36 with a diameter of 30 and a height of –18.
6. Enter SOLCHP with the S options to change:
   (a) radius along the X axis to 20
   (b) radius along the Y axis to 35
   (c) leave the length along the Z axis as –18
   (d) X to end sequence.
7. Rectangular array this cylinder for two rows and two columns, with 100 as the row distance, and 75 as the column distance.
8. Subtract the four cylinders from the box.
9. Create a cylinder at 0,0,0 with diameter 10 and height 18.
10. Use SOLMOVE and enter (a) **L<R>**<R>   – last object
    (b) **TX10,TY10<R>**– motion codes
    (c) **<RETURN>**.
11. Rectangular array this 'hole' for two rows and two columns, the row distance being 80 and the column distance 55.

12. Redrawall.
13. Create a cylinder at 0,0,0 with diameter 20, height 75.
14. Zoom in on the cylinder 'base'
15. Use SOLMOVE with
    (a) **L<R>**<R>   – to pick cylinder
    (b) **E<R>**   – for entity
    (c) **pick bottom circle of cylinder**
    (d) **AF<R>**   – as motion code
    (e) pick long vertical face of cuboid and use N if needed to align MCS icon with this face then**<R>**
    (f) **TX18,TY50,TZ-75<R>**
    (g) **<RETURN>**.
16. Zoom previous – hopefully the cylinder will be aligned along the X axis at the 'middle' of the cuboid.
17. Now enter   SOLHPAT   U
                     SOLHSIZE   4
                     SOLHANGLE   45
18. Use PLINE to draw a polyline
                     from 0,0
                     to 25,30
                     to 0,60
                     to –25,30
                     to close
19. FILLET this polyline with a radius of 6.
20. Now enter SOLMOVE and pick the polyline. Note that shading is added indicating that we have selected a region. Enter the following motion description: **TX37.5,TY20, TZ36<R><R>**.

21. Extrude the repositioned polyline (SOLEXT) by –20 at 0 taper.
22. Redrawall, then subtract the 4 holes, large cylinder and polyline from the box – I used the lower right viewport for this.
23. Your drawing should resemble Fig. 17.1.
24. Save this drawing **SOLID\CASTING**, it will be used again.

Modifying the casting block

We will modify the original casting block by altering the cuboid sizes, changing the horizontal hole and extruding the polyline to cut through the cuboid. We will complete this in stages.

1. Changing the box primitive

Enter SOLCHP**<R>** with the following:
  (a) pick the composite as the solid – primitives back?
  (b) pick the original 'box' as the primitive
  (c) enter S**<R>** with   90 for length along X axis
                 84 for length along Y axis
                 25 for length along Z axis
  (d) enter X to end the sequence.
The composite should be as Fig. 17.2.

2. Changing the horizontal hole

Use SOLCHP and:
  (a) select the composite as the solid
  (b) pick the horizontal cylinder as the primitive
  (c) enter S**<R>** and note the MCS icon orientation
    then   12 for radius along X axis
           5 for radius along Y axis
          90 for length along Z axis
  DO NOT EXIT COMMAND
  (d) enter M**<R>** and move the cylinder from **0,0** by **@15,0**
  (e) enter E**<R>** for evaluate
  (f) X**<R>** to end sequence.

The result of this sequence should be Fig. 17.3.

3. Changing the polyline and copy one of the holes

With SOLCHP:
  (a) pick the composite as the solid
  (b) pick the polyline as the primitive
  (c) enter S**<R>**
  (d) enter N**<R>** in response to the Change shape prompt
  (e) enter –**36<R>** for the extruded height with 0 taper
  (f) E**<R>** to evaluate.
      DO NOT EXIT COMMAND.
  (g) enter N**<R>** until the 'left most hole' is highlighted
  (h) enter I**<R>** – instance for copy
  (i) X**<R>** to end sequence – the hole will be copied onto itself.

4. Making new holes

  (a) Use MOVE with the LAST option, and move the copied hole from 10,10 to 80,20
  (b) with the COPY command, copy this moved hole from 80,20 to 75,75
  (c) redrawall then subtract these two new holes from the composite.

This completes the modifications, and your final casting should resemble Fig. 17.4. Save and plot?

Note

1. The exercise may take some time if you are using a 386 machine, but don't give up on it.
2. The steps in the modification could have been completed using a single SOLCHP command i.e. by not pressing X. I separated the process into stages for convenience.

Task

Can you add dimensions as Fig. 17.5 – handle viewports, layers, etc.

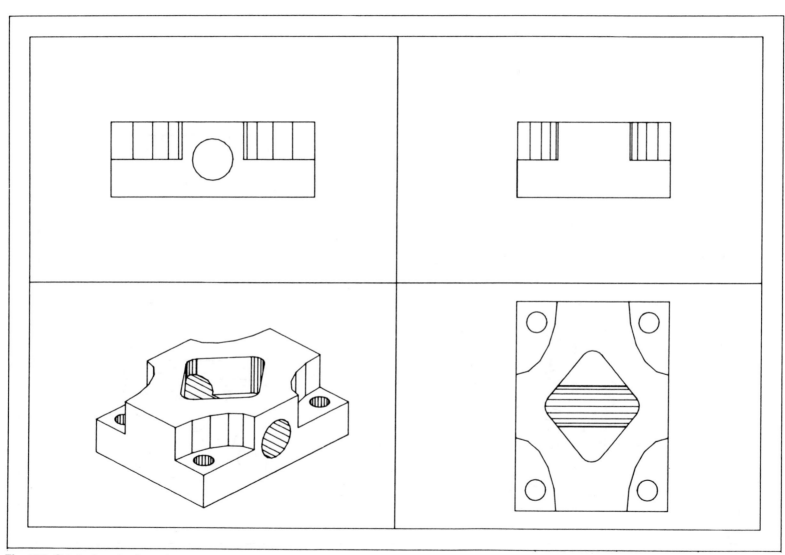

Fig. 17.1. Original casting box plotted with HIDE.

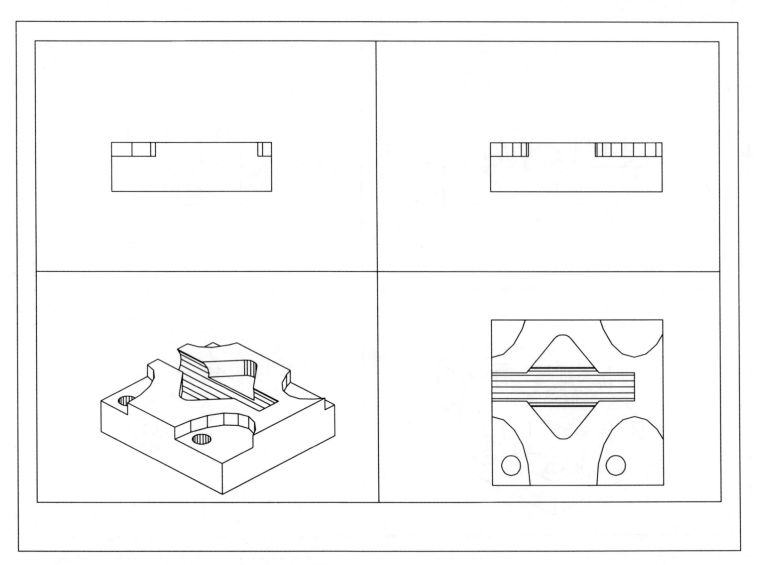

Fig. 17.2. Modified casting block after step 1 plotted with HIDE.

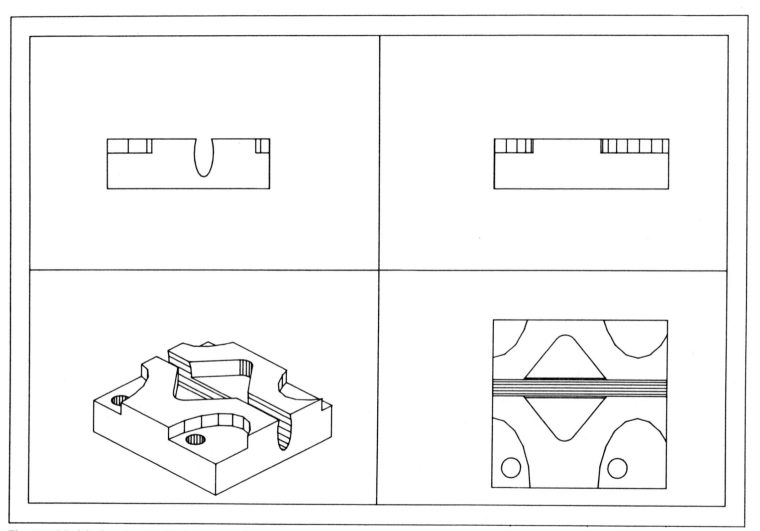

Fig. 17.3. Modified casting block after step 2 plotted with HIDE.

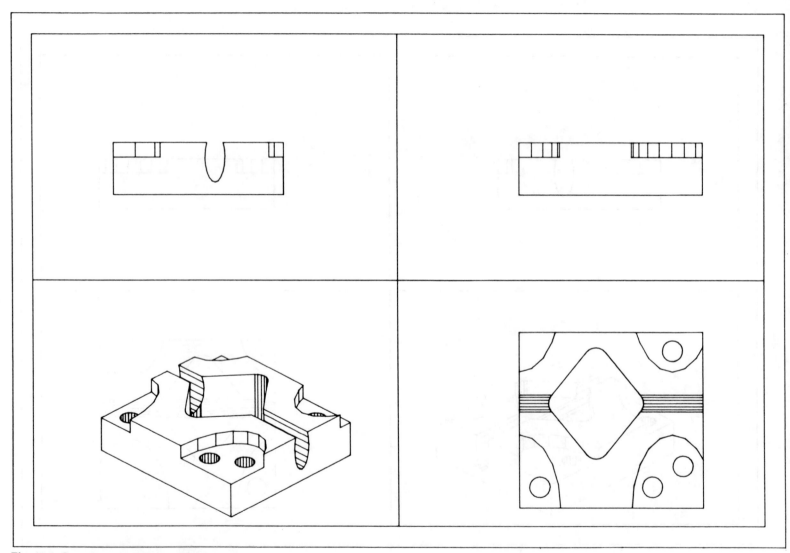

Fig. 17.4. Complete modified casting block plotted with HIDE.

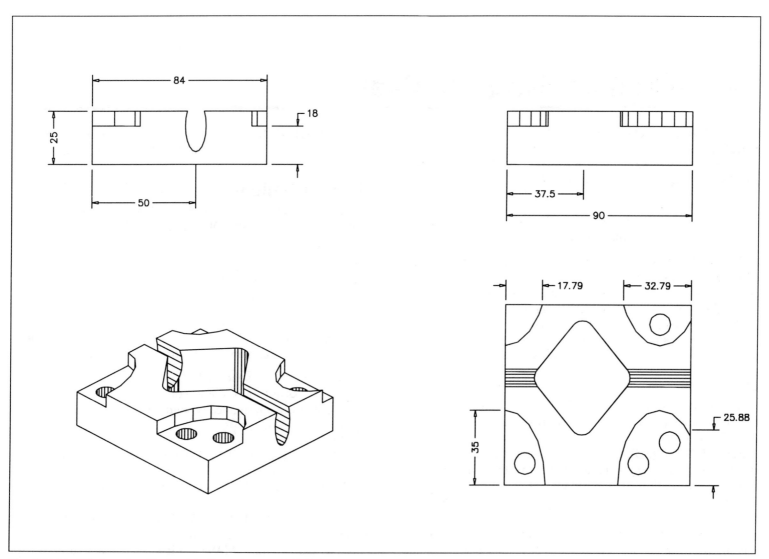

Fig. 17.5. Modified casting block with dimensions with HIDE.

18. Solid modelling variables

We have now completed several solid modelling exercises, and as well as commands, the reader should by now be aware that there are several variables which control how the solid models are drawn. These variables can be changed by:

(a) using the **SOLVAR** command
(b) entering the variable name directly from the keyboard
(c) using the SOLVAR dialogue box.

SOLVAR operates in a similar way to the SETVAR command, i.e. entering **?** at the prompt line will list all variables with their current setting. The solid variables cannot be changed with the SETVAR command. AutoCAD conveniently groups the variables according to their function, and the following is a brief description of them. I should like to point out that several of these variables will probably never be used by the ordinary solid modeller.

Display

SOLWDENS (SOLid Wire DENSity) sets the number of curvature lines (tessellations) in the wire-frame displays, and the number of faces in mesh displays.

SOLDISPLAY (SOLid DISPLAY) sets the display type default for subsequently created solids to either wire-frame or mesh representation. It is normally set to wire-frame mode. The type of display can be changed with SOLWIRE and SOLMESH.

SOLRENDER (SOLid RENDER) determines whether the colours used to display composites will be the colours of the primitives or the colour of the composite.

SOLAXCOL (SOLid AXis COLour) sets the colour of the MCS icon that appears with the SOLMOVE command. Values are 1–8, 1 being red, etc.

Model operations

SOLDELENT (SOLid DELete ENTity) determines when the original entity used with SOLEXT and SOLREV is deleted. The values are:

1. the circle or polyline is not deleted
2. prompt asks question, delete or not
3. automatic deletion.

SOLSOLIDIFY (SOLid SOLIDIFY) determines whether non-solid objects that are eligible to be solidified will be solidified when selected for certain operations. The values are:

1. non-solids are not solidified
2. prompt asks question, solidify or not
3. non-solids are automatically solidified.

Units

SOLLENGTH (SOLid LENGTH units) sets the length units, e.g. millimetres or inches.

SOLAREAU (SOLid AREA Units) sets area units, e.g. square millimetres or square inches.

SOLVOLUME (SOLid VOLUME units) sets volume units, e.g. cubic millimetres or cubic inches.

SOLMASS (SOLid MASS units) sets mass units, e.g. kilogram, pound, ton.

Materials and mass property analysis

SOLMATCURR (SOLid MAterial CURRent) determines the default material that will be assigned to solids. It is **READ ONLY**.

SOLDECOMP (SOLid DECOMPosition direction) sets the decomposition direction used for mass property calculations to either X, Y or Z.

SOLSUBDIV (SOLid SUBDIVision level) determines the number of subdivisions used for mass properties calculations in the SOLMASSP command. There is a formula to work out the number of subdivisions – $(2\wedge s + 1)\wedge 2$, where s is the SOLSUBDIV setting. For example, SOLSUBDIV = 2, 5×5 grid, 25 subdivisions; SOLSUBDIV = 4, 17×17 grid, 289 subdivisions. The higher the value of SOLSUBDIV, the more accurate the calculation, but the longer the time to complete.

2D view extraction

SOLSECTYPE (SOLid SECtion TYPE) determines the contents of blocks created by the SOLSECT command to represent the outline of the section. The values are:

1. block contains lines, arcs, etc.
2. block contains polylines
3. block contains regions.

SOLHPAT (SOLid Hatch PATtern) determines the hatch pattern that will be used when the SOLSECT command is used. It also determines the hatch pattern used with regions. Entering <none> will give an unhatched section/region.

SOLHSIZE (SOLid Hatch SIZE) sets the size of the hatch pattern in sections and regions.

SOLHANGLE (SOLid Hatch ANGLE) sets the hatch pattern angle in sections and regions. The angle is set to the current UCS.

Miscellaneous

SOLAMEVER (SOLid AME VERsion) displays the Release number of AME in use – READ ONLY.

SOLPAGELEN (SOLid PAGE LENgth) sets the number of lines for command display before pausing, e.g. SOLMASSP.

SOLSERVMSG (SOLid SERVer MeSsaGes) specifies how messages will be displayed, e.g.

0 – no messages
1 – error messages only
2 – error messages and beginning/end progress
3 – error messages and full progress messages.

SOLAMECOMP (SOLid AME COMPatibility) controls syntax between AME Release 1 and AME Release 2.

SOLUPGRADE (SOLid UPGRADE) is used for single/double precision depending on what AME is used.

Checking the solid model variables

The solid model variable command can be activated from the:

(a) keyboard with **SOLVAR<R>** – my preference.
(b) screen menu with **MODEL**
                SETUP
                SOLVAR: – lists all variables for selection
(c) menu bar with **Model**
                Setup >
                Variables... – gives the dialogue box.

We will investigate the variables using a previous drawing, so open the **\SOLID\CASTING** drawing just completed in the previous chapter – the original, not the modified drawing and enter **SOLVAR<R>** at the command line:

    AutoCAD prompts   Variable name or ?
    enter   **?<R>**

AutoCAD prompts with a list of the solid variables and their their current values

SOLAMECOMP	AME2	Script compatibility
SOLAMEVER	R2.1	AME release (read only)
SOLAMEAU	sq cm	Area units
SOLAXCOL	3	Solid axes colour
SOLDECOMP	X	Mass property decomposition direction
SOLDELENT	3	Entity deletion
SOLDISPLAY	wire	Display type
SOLHANGLE	45	Hatch angle
SOLHPAT	U	Hatch pattern
SOLHSIZE	4	Hatch size
SOLLENGTH	cm	Length units
SOLMASS	gm	Mass units
SOLMATCURR	MILD_ STEEL	Current material (read only)
SOLPAGELEN	25	Length of text page
SOLRENDER	CSG	Render type
SOLSECTYPE	1	CRoss section representation type
SOLSERVMSG	3	Solid server message display level
SOLSOLIDIFY	3	Automatic solidification
SOLSUBDIV	3	Mass property subdivision level
SOLUPGRADE	0	Upgrade solids to double precision
SOLVOLUME	cu cm	Volume units
SOLWDENS	5	Mesh wire-frame density

The SOLVAR dialogue box

The Solid Variable dialogue box is obtained by selecting the following sequence from the menu bar: **Modify**

Setup >

Variables...

The dialogue box for the **\SOLID\CASTING** drawing being investigated is shown in Fig. 18.1 together with the various optional dialogue boxes which are available. The dialogue box can be used as an alternative to altering the SOLVAR values, allowing the user to 'see' all settings at the one time. The variable name does not have to be remembered as it does if keyboard entry is being used.

❑ **Summary**

Many of the solid variables will never be altered by the user. The most commonly used variables are SOLWDENS, the hatch variables and the model property variables. These are more commonly changed with the Material Property options.

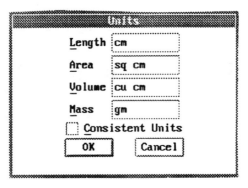

(b) Units option.

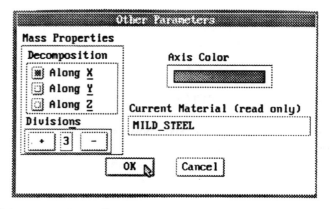

(c) Other parameters option.

(a) Hatch option.

Fig. 18.1. System Variables dialogue box with options.

19. Material properties

Solid modelling is not just about drawing. AutoCAD R12 allows certain 'properties' to be extracted from drawings which are of use to the engineer and designer. These include centroids, moments of inertia, radii of gyration, etc. It is also possible to change the material of a component, thus allowing comparisons to be made regarding the mass of the component. The basic commands which will be used in this chapter are SOLAREA, SOLMASSP and SOLMAT and we will demonstrate the commands using the casting drawing previously created, so open the drawing file **\SOLID\CASTING** – you may still have this drawing loaded if you have continued from the previous chapter.

SOLLIST

This command has been used several times in earlier exercises. It is useful in that it lists the object type and its material. At the command line, enter **SOLLIST<R>**

AutoCAD prompts	`Edge/Face/Tree/<Object>`
enter	**<RETURN>**
AutoCAD prompts	`Select objects`
respond	pick the composite**<R>**
AutoCAD prompts with:	

```
Object type=SUBTRACTION  Handle=??
Component handles:?? and ??
Area=32770.705008  Material=MILD_STEEL
Representation=WIRE-FRAME Render type=CSG
Rigid motion:
```
. . . .

SOLAREA

SOLAREA will calculate the surface area of one or more solids or regions. If more than one solid/region is selected, the total surface area is calculated. The solid is meshed for the purpose of the calculation, and the surface area is based on the sum of all the areas of the elements in the mesh. The higher the value of SOLWDENS, then the more accurate is the final answer, but the time taken for the calculation is longer.

1. From the screen menu select **MODEL**
                                  INQUIRY
                                  SOLAREA

AutoCAD prompts	`Select objects`
respond	pick the composite**<R>**
AutoCAD prompts	`Surface area of solids is` `32770.71 sq cm.`

This is the same as the SOLLIST value – obviously!

2. Now select **MODEL**
                  SETUP
                  ENGINRNG

AutoCAD prompts	`Length unit set to in. Area unit set to sq in. Mass unit set to lbs. Volume unit set to cu in. Display of solids and region is Mesh/<Wire`
enter	**<RETURN>**
AutoCAD prompts	`Wire-frame mesh density (1to12)<5>`

enter   **<RETURN>**

3. Repeat the SOLAREA command and the surface area is displayed as 32770.71 sq in.
4. From the menu bar, select **Model**

         Setup >
         CGS units – default of cm, sq cm, etc.

SOLMASSP

The SOLMASSP command calculates certain mass properties of one or more solids/regions. If several solids are selected, they are considered as one unit for the calculation, and one report is generated. The command has a final prompt which allows the user to save the report in a text file with the extension **.MPR**. This text file can then be dumped to a printer, or imported into other software packages, e.g. spreadsheets/databases, etc. The default for the file prompt is <N>.

1. From the screen menu, select **MODEL**

         INQUIRY
         SOLMASSP

AutoCAD prompts	`Select objects`
respond	pick the composite**<R>**
AutoCAD prompts	`1 solid selected`
	`Calculating mass properties`
	`? of 81 Mass Property`
	`calculation in progress`
then	lists the **REPORT**

2. For the CASTING composite, the complete report is:
Ray projection along *X* axis, level of subdivisions:3

Mass:	1356873 gm
Volume:	172630.1 cu cm (Err:7138.325)
Bounding Box:	X:0 – 75 cm
	Y:0 – 100 cm
	Z:0 – 36 cm

Centroid:	X:37.5 cm	(Err:1.55064)
	Y:47.71341	(Err......
	Z:15.38716	(.........
Moments of inertia:	X:4.676492e+09	gm sq cm
	Y:2.950314e+09	gm sq cm
	Z:6.690144e+09	gm sq cm
Products of inertia:	XY:2.427788e+09	gm sq cm
	YZ:9.871642e+08	gm sq cm
	ZX:7.829404e+09	gm sq cm
Radii of gyration:	X:58.70709 cm	
	Y:46.62991 cm	
	Z:70.21796 cm	

Principal moments (gmsqcm) and *X–Y–Z* directions about centroid:

        I:7.208691e+08 along [2.156203e–16......]
        J:1.69311e+09 along [–1.16686e–16.......]
        K:1.266217e+09 along [1 –2.145289e–16...]

Write to a file<N>

3. At present enter **N<R>** to the last prompt.
4. The materials property report can be viewed in dialogue box form by selecting from the screen menu or menu bar

MODEL	**Model**
INQUIRY	**Inquiry**
DDSOLMSP	**MassProperty...**

The user selects the objects as before. The dialogue box gives the same information as the screen version, see Fig. 19.1.

The report contains information which may be quite meaningless to the user, but if you have ever had to calculate centroid values and moments of inertia on even the most simplest of shapes, you will appreciate the listing given.

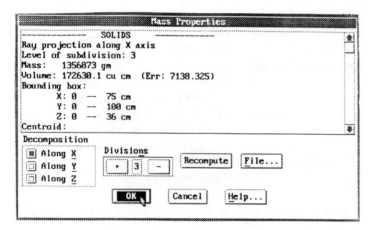

Fig. 19.1. Mass Properties dialogue box.

A brief explanation of the items listed in the report follows:

Ray projection axis: in order to calculate the volume, rays are projected in the direction of the *X, Y* or *Z* axes of the current UCS. The axis is selected with the SOLDECOMP variable.

Level of subdivision: determines the accuracy of the volume calculation, and is set with the SOLSUBDIV variable.

Mass: depends on the object volume and the material density, and as all good engineers know, Mass = Volume × Density.

Volume: is the space occupied by the object.

Centroid: is the centre of mass of the object and assumes that the material is homogeneous. It is often called the centre of gravity. The coordinates of the centroid are given in the report, and the SOLMASSP command *places a point at this location,* although it may be necessary to reset to PDMODE to see it on the actual drawing – a value of 3 is recommended.

Moment of inertia: is used in problems dealing with angular motion and may be considered as a measure of an objects ability to accelerate about a given axis There is a moment of inertia for each axis, this usually being denoted by M*xx*, M*yy* and M*zz*.

Products of inertia: is concerned with the dynamic balancing of rotating objects. For an object to be dynamically balanced about a given axis, the product moment about that axis is 0.

Radius of gyration: is the distance from the rotational axis to the centre of mass. In engineering it is usually denoted by the letter *k*.

Principal moment: is the maximum moment of inertia about an axis through the centroid. This axis is aligned in such a manner that it produces a zero product of inertia.

SOLMAT

This command is used to assign a particular material to a given solid or region. It is also used to keep a record of the property definitions of each material which is available. Material properties are essential for some calculations, e.g. mass = volume × density. Material definitions can be imported into a drawing from a file, or exported from the drawing to a file, and the definitions can be altered by the user at any time. Materials not on file can be added as required and saved for future recall. The file used for this has the name **ACAD.MAT**. Hopefully we will still have the \SOLID\CASTING drawing loaded, so:

1. From the screen menu select **MODEL**
   UTILITY
   SOLMAT:

AutoCAD prompts	`Change/ Edit / LIst / Load/` `New / Remove / SAve/SEt/?/X`
enter	**?<R>**
AutoCAD prompts	with the Select Material File dialogue box as shown in Fig. 19.2
respond	**pick OK** to accept the file ACAD

AutoCAD prompts	`Defined in drawing:`
	`MILD-STEEL`
	`Defined in file:`

`ALUMINUM`	– American spelling
`BRASS`	– soft yellow brass
`BRONZE`	– soft tin bronze
`COPPER`	
`GLASS`	
`HSLA-STL`	– high strength low alloy steel
`LEAD`	
`MILD-STEEL`	
`NICU`	– Monel 400
`STAINLESS-STL`	– austenic stainless steel

AutoCAD prompts	`Change/Edit/..............`
enter	**X<R>** to end this sequence.

2. From the menu bar select **Model**
                    Utility
                    Materials...

AutoCAD prompts	with the Material Browser dialogue box as shown in Fig. 19.3

3. The procedure to change the material is as follows:
   (a) select ALUMINUM from the Materials File list
   (b) pick **Load>** and ALUMINUM is transferred to the other column
   (c) pick ALUMINUM from the Materials in Drawing column, and cancel any other material highlighted in blue – layer idea?
   (d) pick **Set** and ALUMINUM will appear as the Current Material
   (e) pick **Change** and AutoCAD returns to the drawing editor. Simply select to composite then **<RETURN>**
   (f) AutoCAD returns to the dialogue box
   (g) pick **OK**.
4. Use SOLLIST on the composite, and MATERIAL = ALUMINUM.
5. Use SOLMASSP on the composite, and

Mass:	467827.6 gm
Volume:	172630.1 cu cm

6. Other options available with the Materials Browser dialogue box are:
   (a) Edit: gives details of the current material properties (Fig. 19.4)
   (b) New: allows the user to add materials and their properties not stored in the ACAD.MAT file.

These options will be considered in greater detail in the next exercise.

Write to file

1. Change the material of the solid back to MILD_STEEL with:
   (a) enter DDSOLMSP**<R>**
   (b) select MILD_STEEL from the Materials in drawing file and cancel any other, i.e. one blue selection only.
   (c) pick Set
   (d) pick Change and select the composite**<R>**
   (e) pick OK.
2. Check using SOLLIST that the material is MILD_STEEL.
3. Enter SOLMASSP and pick the composite.
4. When the report appears on the screen, enter **Y<R>** to the write to file prompt.
5. AutoCAD responds: with the Create Mass Property File dialogue box as shown in Fig. 19.5.
6. Use the dialogue box by:
   (a) Select Type it
   (b) At the file name prompt enter **\SOLID\CASTING<R>**
   (c) You will be returned to the command prompt.
7. From the menu bar select **File**
   Utilities...
   List files...
   AutoCAD prompts with the `\SOLID` directory and `*.dwg` files ?
8. Change the Pattern name to *.mpr**<R>**
9. The dialogue box should show CASTING as the only *.MPR file in the \SOLID directory.
10. Cancel and exit the utility command.
11. From the screen menu select **UTILITY**
    External Command
    TYPE:
    AutoCAD prompts `file to list:`
    enter **\SOLID\CASTING.MPR<R>**
    AutoCAD prompts `with the CASTING file report`

12. If you have a printer connected to your system, you can obtain a hard copy of this report by:
    (a) enter TYPE at the command prompt
    (b) entering **\SOLID\CASTING.MPR>PRN** at the file name prompt.

This completes the materials property section. I have laboured some of the points as this is an important area of solid modelling. Our next exercise will investigate in greater detail the mass properties of a solid we will create.

❏ *Summary*

1. Material properties are used for calculations.
2. These properties are contained in the ACAD.MAT file and the user can add materials to this file as required.
3. The commands used with the material properties are SOLAREA, SOLMASSP and SOLMAT.
4. Dialogue boxes are available if required.

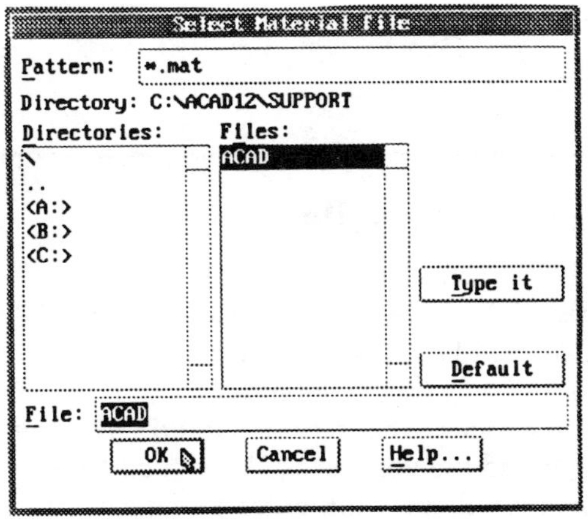

Fig. 19.2. Select Material dialogue box.

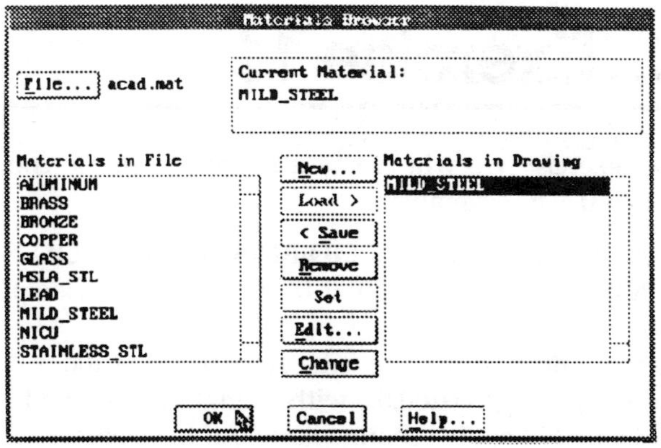

Fig. 19.3. Materials Browser dialogue box.

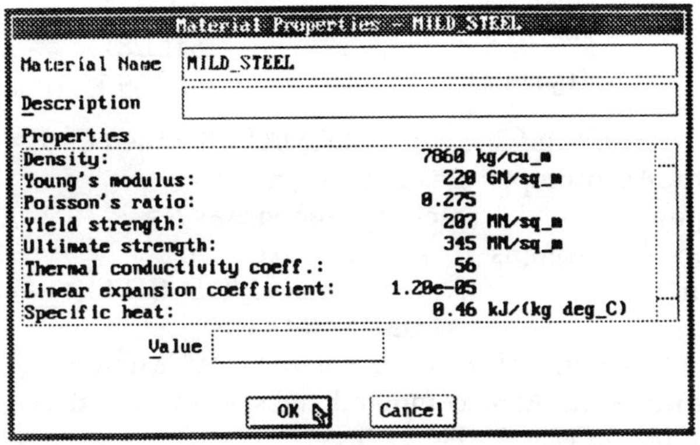

Fig. 19.4. Materials Browser (Edit) dialogue box.

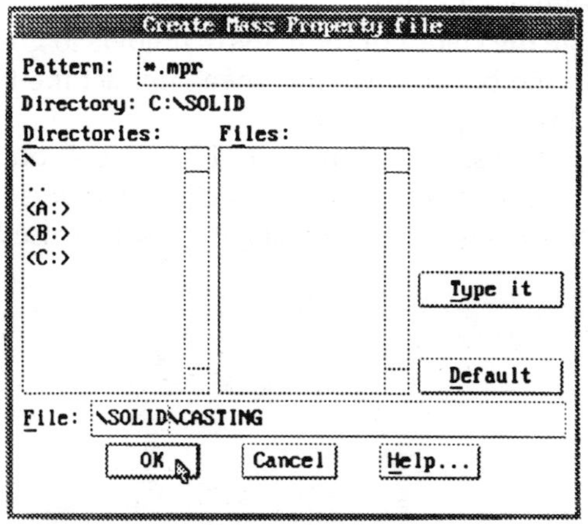

Fig. 19.5. Create Mass Property File dialogue box.

20. Exercise 11

In this exercise we will create a relatively simple composite, and then use it for analysis.

1. Begin a new drawing **SOLID\BLOCKAN=\SOLID\SOLA3**.
2. ZOOM C about 60,50,50 at 0.6XP in the 3D viewport and 1XP in the other three.
3. Use SOLBOX to create a cube at 0,0,0 of side 100.
4. Create a wedge at 100,0,0 with a length of 40, a width of 100 and a height of 70 – change its colour to yellow.
5. Create two cylinders:
   (a) at 50,0,50 with diameter 50 and use the centre option with @0,100,0
   (b) at 0,50,50 with diameter 50 and use the centre option with @140<0<0.
6. Change the colour of these two cylinders to green.
7. Union the box and wedge, then subtract the two cylinders from the composite.
8. The composite is shown in Fig. 20.1 – note the interpenetration effect of the intersecting cylinders.
9. Save the composite at this stage – it will be used later.

Analysis of composite

1. Use SOLLIST to check material – MILD_STEEL?
2. With SOLAREA, check the surface area is 83076.78 sq cm.
3. Set variables SOLDECOMP to X axis and SOLSUBDIV to 3 – these will probably be the defaults.
4. Enter SOLMASSP. The report should give the mass as 6413281 gm and the volume as 815939.1 cu cm?
5. At the write to file prompt, enter **Y** and save the file using the name **SOLID\BLOCK_1** – Fig. 20.2.
6. Using MODEL then SETUP, change the units, and complete the following table:

BLOCKAN MILD_STEEL

	ENGINRNG	IMPERIAL	CGS	SI
Surface area	83076.78	....	83076.78	....
Mass	231694.5	....	6413281	....
Volume	815939.1	....	815939.1	....

7. At the command line, enter SOLMAT**<R>** then C**<R>**
   AutoCAD prompts   Select objects
   respond   **pick the composite<R>**
   AutoCAD prompts   1 solid selected
   then   New material<MILD_STEEL>
   enter   **ALUMINUM<R>**
8. Now use SOLLSIT to check material – ALUMINUM?
9. Use the SOLMASSP command and save the report as \SOLID\BLOCK_2 – Fig. 20.3.

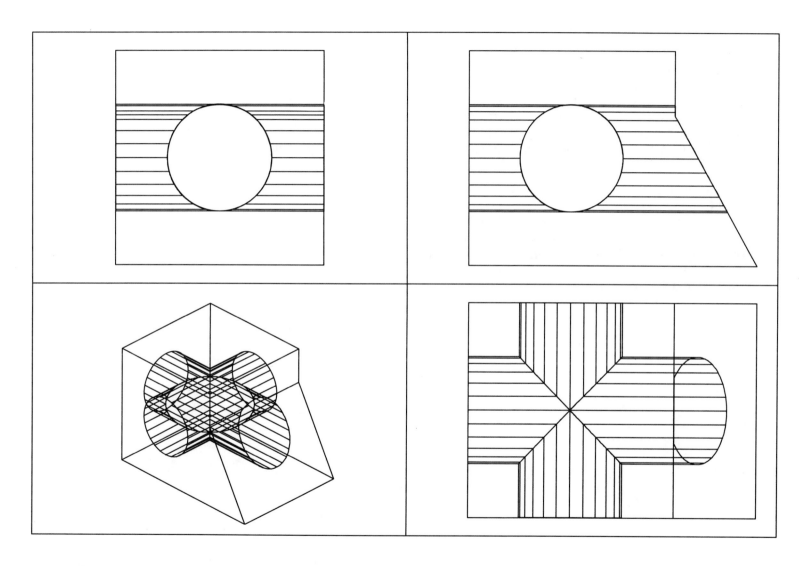

Fig. 20.1. Block for analysis plotted without HIDE.

10. With the SOLMAT command, and with CGS units, complete the table overleaf for the different materials.

BLOCKAN CGS units

	MILDSTEEL	ALUMINUM	BRONZE	LEAD	STAINLESS
Surface area	83076.78	83076.78	....	......	....
Mass	6413281	2211195	7240644	....	....
Volume	815939.1	....	....	....	....
File name	BLOCK–1	BLOCK–2	BLOCK–3	BLOCK–4	BLOCK–5

11. The complete report for the BRONZE material is given in Fig. 20.4.
12. Before leaving this chapter, restore the MILD_STEEL material and then MOVE the composite from 0,0,0 to 100,0,0 and recalculate the mass properties. Save the report as BLOCKMOD – Fig. 20.5, and compare the results with those from BLOCK_1.

```
Ray projection along X axis, level of subdivision: 3.
Mass:            6413281 gm
Volume:          815939.1 cu cm   (Err: 29306.07)

Bounding box:            X: 0   --   140 cm
                         Y: 0   --   100 cm
                         Z: 0   --   100 cm

Centroid:                X: 59.6222   cm     (Err: 2.882734)
                         Y: 50.14488  cm     (Err: 4.54186)
                         Z: 45.16848  cm     (Err: 4.138924)

Moments of inertia:      X: 4.20612e+10   gm sq cm (Err: 3.854576e+09)
                         Y: 5.144858e+10  gm sq cm (Err: 2.103596e+09)
                         Z: 5.35018e+10   gm sq cm (Err: 3.648584e+09)
Products of inertia:    XY: 1.913817e+10  gm sq cm (Err: 1.864849e+09)
                        YZ: 1.459094e+10  gm sq cm (Err: 1.947254e+09)
                        ZX: 1.570662e+10  gm sq cm (Err: 1.225792e+09)

Radii of gyration:       X: 80.98428   cm
                         Y: 89.5667    cm
                         Z: 91.33644   cm

Principal moments(gm sq cm) and X-Y-Z directions about centroid:
                         I: 1.19259e+10 along [0.8609277 -0.01758364
                                                         -0.5084234]
                         J: 1.558452e+10 along [0.1960854 -0.9107106
                                                          0.3635335]
                         K: 1.548406e+10 along [-0.4694189 -0.4126705
                                                          -0.7806081]
```

Fig. 20.2. BLOCK_1 report for BLOCKAN with MILD_STEEL.

```
Ray projection along X axis, level of subdivision: 3.
Mass:               2211195 gm
Volume:             815939.1 cu cm   (Err: 29306.07)

Bounding box:            X: 0  --  140 cm
                        Y: 0  --  100 cm
                        Z: 0  --  100 cm

Centroid:               X: 59.6222   cm     (Err: 2.882734)
                        Y: 50.14488  cm     (Err: 4.54186)
                        Z: 45.16848  cm     (Err: 4.138924)

Moments of inertia:    X: 1.450201e+10  gm sq cm (Err: 1.328995e+09)
                       Y: 1.773863e+10  gm sq cm (Err: 7.252856e+08)
                       Z: 1.844655e+10  gm sq cm (Err: 1.257972e+09)
Products of inertia: XY: 6.598529e+09  gm sq cm (Err: 6.429696e+08)
                     YZ: 5.03072e+09  gm sq cm (Err: 6.713815e+08)
                     ZX: 5.415389e+09  gm sq cm (Err: 4.226333e+08)

Radii of gyration:     X: 80.98428   cm
                       Y: 89.5667    cm
                       Z: 91.33644   cm

Principal moments(gm sq cm) and X-Y-Z directions about centroid:
                       I: 4.111858e+09 along [0.8609277 -0.01758364
                                                       -0.5084234]
                       J: 5.37329e+09 along [0.1960854 -0.9107106
                                                        0.3635335]
                       K: 5.338654e+09 along [-0.4694189 -0.4126705
                                                        -0.7806081]
```

Fig. 20.3. BLOCK_2 report for BLOCKAN with ALUMINUM.

```
Ray projection along X axis, level of subdivision: 3.
Mass:              7240644 gm
Volume:            815939.1 cu cm   (Err: 29306.07)

Bounding box:          X: 0  --  140 cm
                       Y: 0  --  100 cm
                       Z: 0  --  100 cm

Centroid:              X: 59.6222   cm    (Err: 2.882734)
                       Y: 50.14488  cm    (Err: 4.54186)
                       Z: 45.16848  cm    (Err: 4.138924)

Moments of inertia:    X: 4.748742e+10  gm sq cm (Err: 4.351846e+09)
                       Y: 5.808585e+10  gm sq cm (Err: 2.374976e+09)
                       Z: 6.040395e+10  gm sq cm (Err: 4.119279e+09)
Products of inertia:  XY: 2.160714e+10  gm sq cm (Err: 2.105429e+09)
                      YZ: 1.647328e+10  gm sq cm (Err: 2.198465e+09)
                      ZX: 1.77329e+10   gm sq cm (Err: 1.383929e+09)

Radii of gyration:     X: 80.98428   cm
                       Y: 89.5667    cm
                       Z: 91.33644   cm

Principal moments(gm sq cm) and X-Y-Z directions about centroid:
                       I: 1.346443e+10  along [0.8609277 -0.01758364
                                                          -0.5084234]
                       J: 1.759504e+10  along [0.1960854 -0.9107106
                                                           0.3635335]
                       K: 1.748162e+10  along [-0.4694189 -0.4126705
                                                          -0.7806081]
```

Fig. 20.4. BLOCK_3 report for BLOCKAN with BRONZE.

```
Ray projection along X axis, level of subdivision: 3.
Mass:            6413281 gm
Volume:          815939.1 cu cm  (Err: 29306.07)

Bounding box:        X: 100  --   240 cm
                     Y: 0    --   100 cm
                     Z: 0    --   100 cm

Centroid:            X: 159.6222   cm   (Err: 6.386955)
                     Y: 50.14488   cm   (Err: 4.54186)
                     Z: 45.16848   cm   (Err: 4.138924)

Moments of inertia:  X: 4.20612e+10   gm sq cm (Err: 3.854576e+09)
                     Y: 1.920561e+11  gm sq cm (Err: 6.84476e+09)
                     Z: 1.941094e+11  gm sq cm (Err: 8.696567e+09)
Products of inertia: XY: 5.129748e+10  gm sq cm (Err· 4.751585e+09)
                     YZ: 1.459094e+10  gm sq cm '        1.947254e+09)
                     ZX: 4.467444e+10  gm sc          3.876656e+09)

Radii of gyration:   X: 80.98428   cm
                     Y: 173.051    cm
                     Z: 173.9735   cm

Principal moments(gm sq cm) and X-Y-Z direc ons        troid:
                     I: 1.19259e+10 along [  ˊ        ˏ.01758364
                                                    -0.5084234]
                     J: 1.558452e+10 along [  ˌ960854 -0.9107106
                                                    0.3635335]
                     K: 1.548406e+10 along [-0.4694189 -0.4126705
                                                    -0.7806081]
```

Fig. 20.5. BLOCKMOD report for BLOCKAN with MILD_STEEL repositioned.

Entering our own Material Values

1. Re-open the drawing file that was saved earlier in the exercise, i.e. **\SOLID\BLOCKAN** and from the screen menu select **MODEL**
   UTILITY
   DDSOLMAT

   AutoCAD prompts: with the Materials Browser dialogue box.
2. From the dialogue box select **New...** .

   AutoCAD prompts   with the New Materials dialogue box
                      – empty?
   (a) in the Material Name box, enter: **BOBSMAT**
   (b) in the Description box, enter: **A newly discovered rust free 'steel'**
   (c) pick **Density** and in the Value box enter: **9650<R>**
   (d) pick **Young's modulus** and in the Value box enter: **450<R>**
   (e) complete the entries for the other properties using the figures given in Fig. 20.6.

Note: (i)   the linear expansion entry is **figure/1e6**
         (ii) if the entry is 0.2, the value is **2.00e-01**
   (f) pick OK when all entries are complete.
3. Now make the material in the drawing **BOBSMAT**.
4. Pick **Set**.
5. Pick **Change** and pick the composite.
6. Pick OK.
7. SOLLIST and check material – BOBSMAT?
8. DDSOLMASSP and save to file – Fig. 20.7.

This completes the material properties' exercise. AutoCAD allows different material properties to be entered and saved. The mass property calculations are useful, fast and the ability to change materials allows comparisons to be made with a few keystrokes. The printed report is also useful, and being a text file, can be imported into spreadsheets and word-processors. The reports listed in here were produced in WordPerfect 5.1.

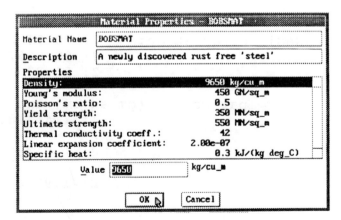

Fig. 20.6. Materials Browser for new material BOBSMAT.

```
Ray projection along X axis, level of subdivision: 3.
Mass:            7873812 gm
Volume:          815939.1 cu cm  (Err: 29306.07)

Bounding box:          X: 0  --  140 cm
                       Y: 0  --  100 cm
                       Z: 0  --  100 cm

Centroid:              X: 59.6222   cm    (Err: 2.882734)
                       Y: 50.14488  cm    (Err: 4.54186)
                       Z: 45.16848  cm    (Err: 4.138924)

Moments of inertia:    X: 5.164003e+10  gm sq cm (Err: 4.7324e+09)
                       Y: 6.316525e+10  gm sq cm (Err: 2.582659e+09)
                       Z: 6.568606e+10  gm sq cm (Err: 4.479496e+09)
Products of inertia:  XY: 2.34966e+10   gm sq cm (Err: 2.289541e+09)
                      YZ: 1.791381e+10  gm sq cm (Err: 2.390713e+09)
                      ZX: 1.928358e+10  gm sq cm (Err: 1.504949e+09)

Radii of gyration:     X: 80.98428  cm
                       Y: 89.5667   cm
                       Z: 91.33644  cm

Principal moments(gm sq cm) and X-Y-Z directions about centroid:
                       I: 1.464185e+10 along [0.8609277 -0.01758364
                                                        -0.5084234]
                       J: 1.913367e+10 along [0.1960854 -0.9107106
                                                         0.3635335]
                       K: 1.901033e+10 along [-0.4694189 -0.4126705
                                                         -0.7806081]
```

Fig. 20.7. BLOCKBOB report for new material BOBSMAT.

21. Exercise 12: extracting sections

This exercise will introduce the concept of extracting hatching details from composites. We will use the block created (and saved) in the previous chapter.

1. Begin a new drawing \SOLID\DRSECT=\SOLID\BLOCKAN.
2. In model space, use the SOLCHP command and alter:
   (a) the radius of the shorter cylinder to 15 along both the X and Y axes.
   (b) the radius of the longer cylinder to 10 along the X axis and 30 along the Y axis.
   (c) the length along the Z axis is to remain the same.
3. With the upper right viewport active, restore UCS LEFT.
4. Position the UCS icon at the origin in **ALL** viewports.
5. Use the UCS command with the (O)rigin option to move the UCS from (0,0,0) to (0,0,–50)
6. Save this UCS setting as SECT1.
7. For the four PV-? current viewport layers:
   (a) thaw each layer
   (b) change the colour of the four layers to CYAN
   (c) make the upper right viewport layer current, the author's was PV-7 – handle numbers needed.
   Remember: PV-7 On...N – Current VP layer thawed
   PV-8 ON..FN – Current VP layer frozen
8. Set the hatch variables   SOLHPAT – ANSI31
                                SOLHSIZE – 2
                                SOLHANGLE – 0
9. Enter **SOLSECT** <R>

AutoCAD prompts   `Select objects`
respond   **pick the composite<R>**
AutoCAD prompts   `1 solid selected`
then   `Sectioning plane by Entity/`
                  `Last....`
enter   *XY<R>*
AutoCAD prompts   `Point on XY plane <0,0,0>`
enter   **0,0,0<R>**

10. The block should then be hatched in cyan as Fig. 21.1.
11. Erase all cyan entities from the screen – erase twice?
12. Restore WCS.
13. With lower left viewport active, make PV-9 the current layer remembering the correct handle number.
14. Using the (O) option of the UCS command, reposition the origin at (0,0,40).
15. Repeat the SOLSECT command:
    (a) picking the composite
    (b) entering XY
    (c) entering 0,0,0.
16. Result of this sectioning is Fig. 21.2.

This has been an introduction to extracting sectional details from composites, and will be further expanded in the next chapter. Before leaving this chapter think about these:
   (a) why was XY entered at the SOLSECT prompt?
   (b) what would entering YZ have given?
   (c) how can the hatching only be displayed in one viewport?

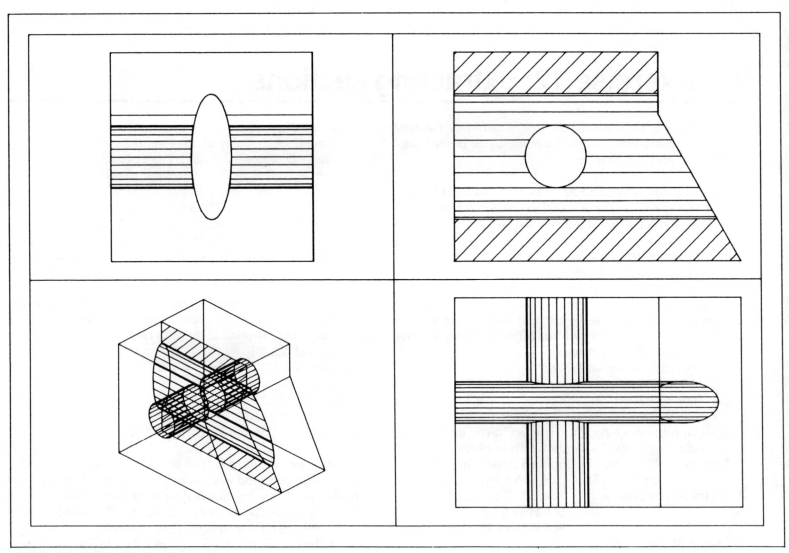

Fig. 21.1. BLOCKAN after hatching with UCS at SECT1 without HIDE.

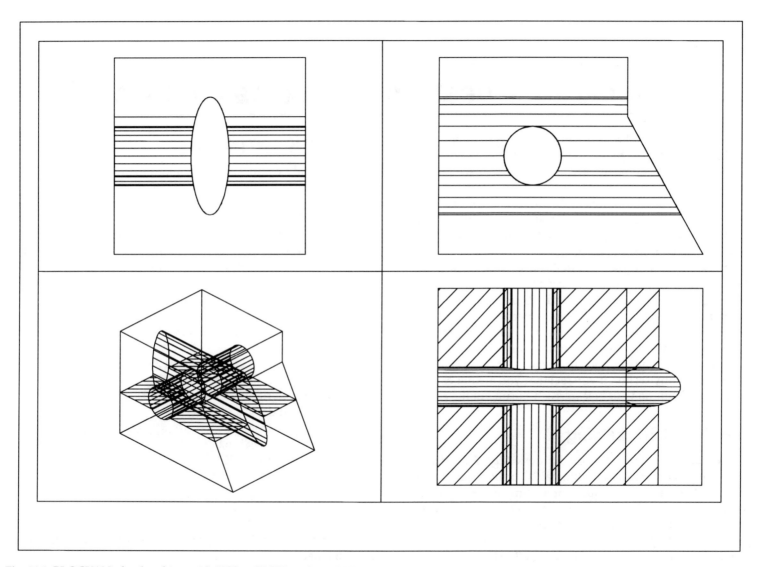

Fig. 21.2 BLOCKAN after hatching with UCS at SECT2 without HIDE.

22. Exercise 13: a desk tidy as a detail drawing

This exercise is complex and will involve you in a lot of layer control, especially the PV and PH layers. It is essential that you know the handle numbers of your viewports. I would advice you to sketch out the four viewports on a piece of scrap paper and enter the handle numbers. In the exercise I will use my viewport handles, but will constantly refer to the viewport position, e.g. upper left, lower right, etc. to assist you in determining your handle numbers.

The exercise is interesting in that it will extract profile details from a composite solid to produce a 'traditional 2D' orthogonal drawing having a section view and a true shape drawn as an auxiliary view. A 3D view will also be obtained.

The steps in the exercise are fully given, but only expanded prompts are given for 'newish' commands.

I have divided the exercise into sections, and you can give up at the end of any section if you feel that you are not capable of continuing – but do your best to reach the end of the exercise.

Traditional 2D drawing

Figure 22.1 gives the sizes for the desk tidy. As a diversion draw the given views as a traditional 2D orthogonal drawing. I think that you will find it quite interesting, especially with the three holes on the sloped surface. Add an isometric view which should not present you with too many problems. There are two sizes which must be noted from this drawing, these being:
1. the slope surface short end length – 18.68154169
2. the angle of the slope surface to the horizontal – 15.524111.

The basic composite

The desk tidy composite is made originally from an extruded polyline and then primitives are added and subtracted as required.

1. Begin a new drawing \SOLID\TIDY=\SOLID\SOLA3 which should put you in model space in the lower left viewport with SOLIDS as the current layer (red).
2. In paper space, use the STRETCH command and window the yellow viewport crossing in the centre of the screen. Enter 0,0 as the base point and −25,25 as the second point of displacement. This will have the effect of 'shrinking' the top viewports.
3. In model space ZOOM C about 75,38,12 at 0.75XP in the 3D viewport and 1XP in the other three viewports.
4. In the top right viewport, restore the UCS LEFT. Use PLINE and draw a polyline
      from 0,0
      to @156,0
      to @0,15
      to @−132,0
      to @0,10
      to @−24,0
      to close
5. Extrude this polyline by 85 in the negative Z direction with 0 taper.
6. Restore WCS with the lower left viewport active.

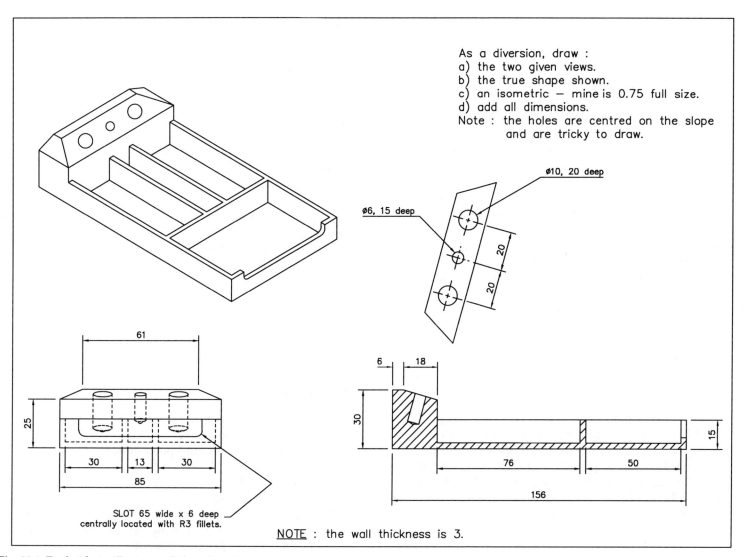

As a diversion, draw :
a) the two given views.
b) the true shape shown.
c) an isometric — mine is 0.75 full size.
d) add all dimensions.
Note : the holes are centred on the slope and are tricky to draw.

Ø10, 20 deep

Ø6, 15 deep

20

20

61

25

30   13   30

85

SLOT 65 wide x 6 deep
centrally located with R3 fillets.

6   18

30

15

76   50

156

NOTE : the wall thickness is 3.

Fig. 22.1. Desk tidy in 2D giving all the relevant sizes.

7. Create a box with SOLBOX at 0,12,25 with length 6, width 61 and height 5.
8. Create three wedges as follows:
   (a) at 6,12,25 with length 18, width 61 and height 5
   (b) at 6,12,25 with length 12, width –6 and height 5. Rotate this wedge about the point 6,12,25 by –90°
   (c) at 6,73,25 with length 12, width 6 and height 5. Rotate this wedge 90° about the point 6,73,25.
9. Now union these three wedge and the box with the extrusion.
10. Create two wedges:
    (a) at 6,0,25 with length 18, width 12 and height 5
    (b) at 24,12,25 with length 12, width –18 and height 5. Rotate this wedge about the point 24,12,25 by –90°.
11. Intersect these two wedges with SOLINT.
12. Create another two wedges:
    (a) at 6,73,25 with length 18, width 12 and height 5
    (b) at 24,73,25 with length 12, width 18 and height 5 and rotate it by 90° about the point 24,73,25.
13. Intersect these two wedges and then union the four intersected wedges with the composite.

The compartments

This will be achieved with four boxes which will be subtracted from the composite.

1. Still in the lower left viewport with WCS, use SOLBOX:
   (a) at 153,3,3 with length –50, width 79 and height 20
   (b) at 100,3,3 with length –76, width 30 and height 20
   (c) at 100,36,3 with length –76, width 13 and height 20
   (d) at 100,52,3 with length –76, width 30 and height 20.
2. Change the colour of these four boxes to yellow and then subtract them from the composite.
3. At this stage I would recommend you saving your work.
4. Also try SOLMESH, HIDE, SHADE and return to wire-frame – REGEN and SOLWIRE.

The front cut out

The front cut out will be obtained with an extruded polyline.
1. In lower left viewport make a new UCS on the front (red) face using the 3 point option with
   (a) origin at 156,0,0
   (b) X axis at 156,85,0
   (c) Y axis at 156,0,15.
2. Icon should be at origin point, if not then UCSICON – OR.
3. Save this UCS setting as NEWFRONT – it may be used again.
4. Draw a polyline
   > from 10,15
   > to @0,–3
   > arc to @3,–3
   > line to @59,0
   > arc to @3,3
   > line to @0,3
   > to close
5. Change the colour of this polyline to blue.
6. Extrude the polyline by –3 with 0 taper.
7. Subtract the blue extrusion from the composite.

The holes

1. In the lower left viewport restore WCS.
2. Zoom in on the 'sloped face'
3. Make a new UCS with the 3 point option
   (a) origin at 24,42.5,25
   (b) X axis at 24,85,25
   (c) Y axis at 6,42.5,30.
4. The UCS should be aligned at the midpoint of the bottom edge of the slope and be 'lying on the slope' and at the origin. UCSICON ORigin if it is not.
5. Save this UCS as SLOPE.

6. Zoom previous.
7. Create a cylinder at 0,9.34077,0 with diameter 6 and a height of −12. The 9.34077 comes from the slope dimension in the first drawing – remember the sizes you had to note?
8. Copy this cylinder from 0,9.34077 to @20,0
9. Now change the colour of the first cylinder to cyan and the copied cylinder to green.
10. Zoom tightly in on the bottom end of the cyan cylinder.
11. Create a cone at 0,9.34077,−12 with diameter 6 and height −3 and change its colour to cyan.
12. Union the cyan cone and cylinder.
13. Zoom previous, then zoom in on the green cylinder.
14. Use SOLCHP with the S option to change the cylinder radius to 5 and the length to −17.
15. Create a cone at 20,9.34077,−17 with diameter 10 and height −3 and change its colour to green.
16. Union the green cone and cylinder.
17. Zoom back out, and copy the cone/cylinder from 20,9.34077,0 by @−40,0.
18. Subtract the three 'holes' from the composite.
19. Restore WCS and save your drawing at this stage – it should resemble Fig. 22.2.
20. SOLMESH etc and return to wire-frame.
21. Now that the composite is complete, use your knowledge of materials properties to list the surface area, mass and volume for two materials, MILD_STEEL and BRASS. My results were:

	MILD STEEL	*BRASS*
Surface area	47279.61 sq cm	47279.61 sq cm
Mass	889001.2 gm	937713.6.gm
Volume	110710 cu cm	110710 cu cm

The auxiliary feature extraction

To extract the true shape of the sloped surface as an auxiliary we will use the SOLFEAT command, and move the extracted feature in a later chapter.

1. With the lower right viewport active, restore UCS SLOPE. The icon should be at the origin, if it is not then UCSICON ORigin.
2. Enter PLAN<R> to give a plan view of the slope. This will alter the orientation of the view.
3. Enter **DVIEW** and pick the composite

AutoCAD prompts	CAmera/TWist/............
enter	**TW<R>**
AutoCAD prompts	New view twist
enter	**74.475889<R>** – from the 15.524111 value
AutoCAD prompts	CAmera..................
enter	**X<R>**

4. Restore WCS and ZOOM C about 75,38,12 at 1XP
5. You should now be looking 'straight down' the holes.
6. Enter **VIEW<R>** and save the view as SLOPE.
7. Make PV-A the current layer (watch the handles)
8. Extract the slope features with **SOLFEAT<R>**

AutoCAD prompts	Edge/<Face>
enter	**F<R>**
AutoCAD prompts	All/<Select>
enter	**S<R>**
AutoCAD prompts	pick a face
respond	**pick the bottom sloped line of the shape** and use the **N** option until the slope outline is highlighted then**<R><R>**
AutoCAD prompts	by highlighting the slope face in a green–black dotted linetype.

9. Now redrawall.

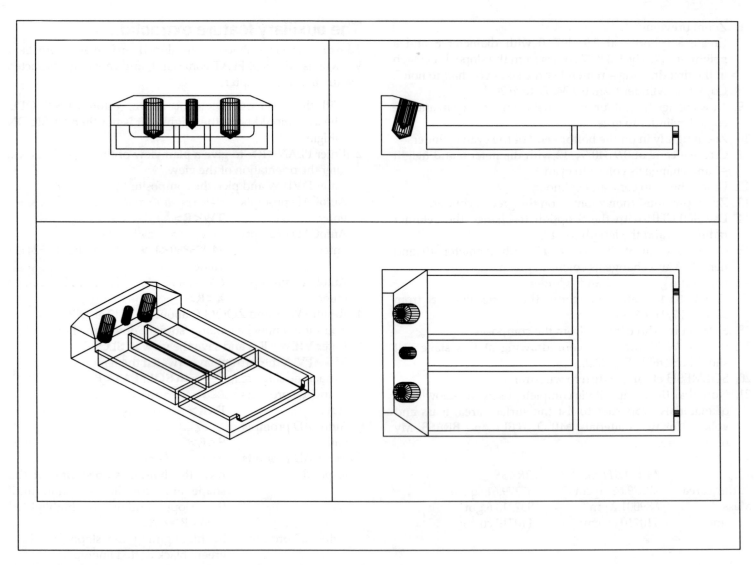

Fig. 22.2. Desk tidy – basic composite plotted without HIDE.

The section extraction

1. In the upper right viewport restore UCS LEFT.
2. Icon at origin – it has to be.
3. Using the UCS **O** option, position the UCS at the point 0,0, –42.5 and save it as SECT.
4. Make PV-7 the current layer and freeze all other PV layers.
5. Set the following variables: SOLHPAT to U, SOLHSIZE to 3 and SOLHANGLE to 45.
6. Enter **SOLSECT** and select the composite. Enter **XY** to the plane prompt and **0,0,0** to the point prompt.
7. Hopefully a green section will be drawn only in the top right viewport as shown inFig. 22.3.
8. Save again at this stage.

Extracting the profiles

1. With the lower left viewport active, return to WCS.
2. ZOOM at 0.75XP to extract a profile the same size as the view. Note: ZOOM 0.75XP **NOT** ZOOM C 0.75XP.
3. Enter **SOLPROF** and pick the composite

AutoCAD prompts	`Display hidden profile lines on sep. layers`
enter	**Y<R>**
AutoCAD prompts	`Project profile lines onto a plane`
enter	**Y<R>**
AutoCAD prompts	`Delete tangential edges`
enter	**Y<R>**
AutoCAD prompts	by adding 'rust coloured' hidden detail

4. Repeat steps 2 and 3 in the upper left viewport but ZOOM 1XP.

Tidying up the drawing

1. Restore WCS if needed.
2. Freeze the layer SOLIDS to leave the green profiles.
3. In the lower left viewport erase the 'hidden lines' – zoom may be necessary, but remember to zoom out.
4. In the upper right viewport, restore UCS FRONT.
5. Make PH-8 the current layer and freeze PV-8.
6. Explode the 'hidden line' block (turns black).
7. Use CHPROP and change the colour of the lines to 'Bylayer'.
8. Now carefully erase all hidden lines which would coincide with outlines. This will mean erase/redraw a few times. This will require you to know what hidden lines are extra to the view.
9. Thaw layer PV-8.
10. In the upper right viewport make PV-7 the current layer.
11. Add the five lines needed to complete the view, using ORTHO and TRIM to assist you.
12. Your drawing should now be as shown in Fig. 22.4.
13. Once again save.

Moving the viewports

1. In paper space interchange the two left viewports. I'll let you figure this out for yourself, but OSNAP END is a hint – refer to Fig. 22.5. for guidance.
2. Move the upper right viewport to the bottom.
3. To move the auxiliary viewport, enter MOVE and pick the yellow viewport border.
   (a) at the base prompt – OSNAP END and pick the lower right corner of the 'true shape'
   (b) at the second point prompt – OSNAP END and pick a coincident point as shown in Fig. 22.5.
4. Now move the 'true shape' viewport form 0,0 by **@62<74.475889**
5. The result of this should be as shown in Fig. 22.6.

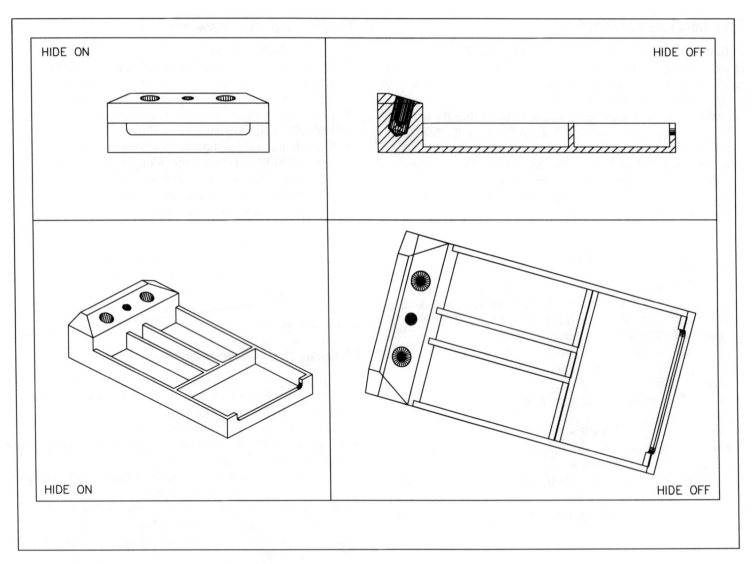

HIDE ON

HIDE OFF

HIDE ON

HIDE OFF

Fig. 22.3. Desk tidy – after SOLFEAT and SOLSECT.

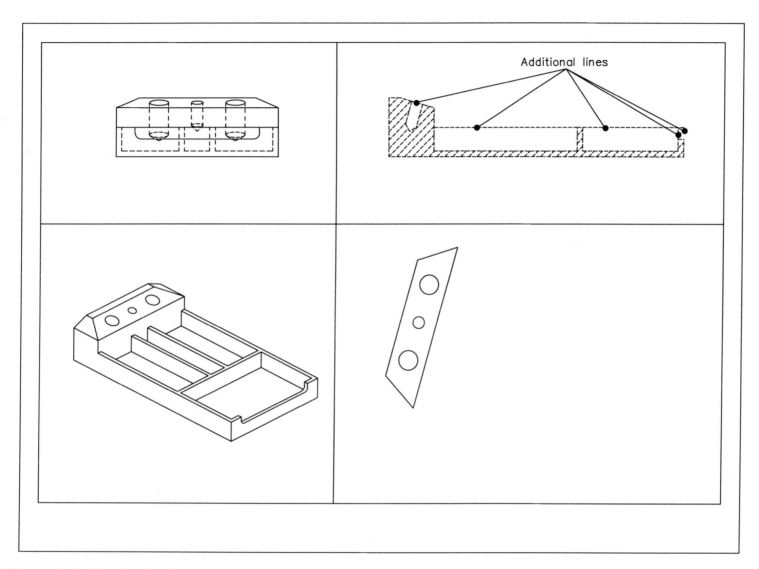

Fig. 22.4. Desk tidy – after SOLPROF and cleaning up.

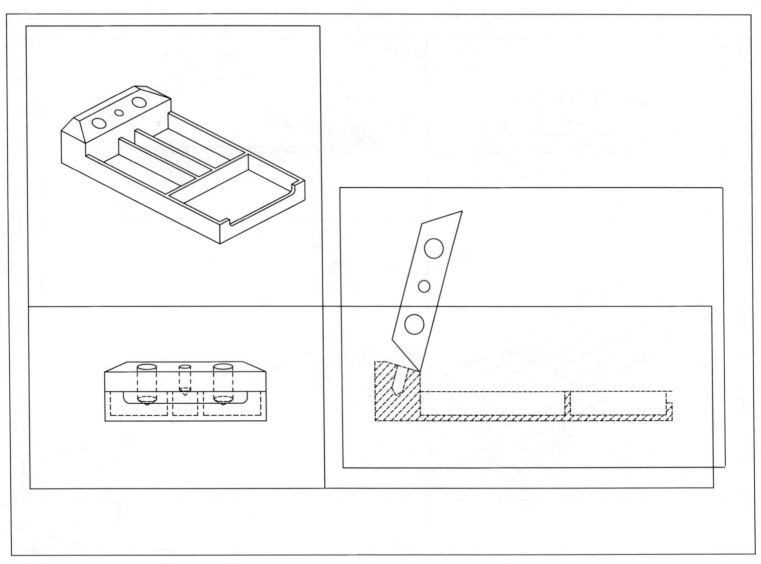

Fig. 22.5. Desk tidy – after moving the viewports.

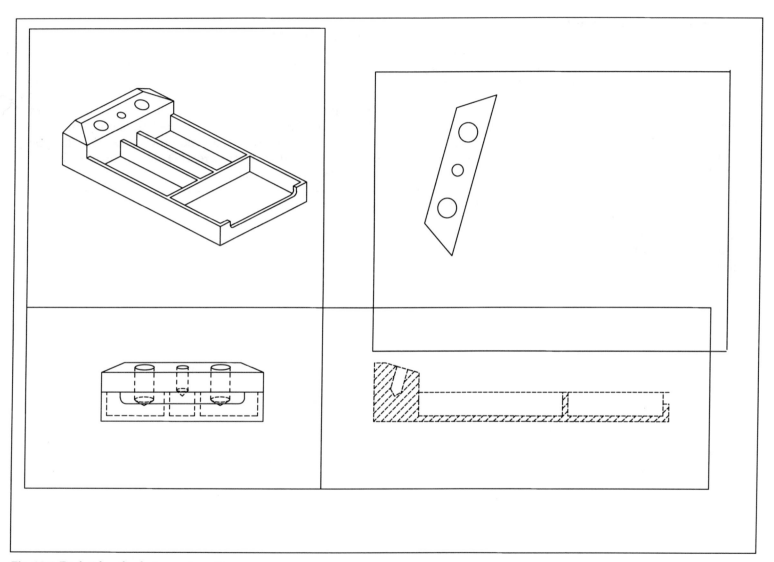

Fig. 22.6. Desk tidy – final viewport positions.

Dimensioning the drawing

We will add dimensions to our drawing. Remember that to dimension it is necessary to have the active viewport DIM layer current and the other DIM layers frozen. The correct UCS must be restored, e.g. LEFT, FRONT.

1. Dimension the two main views as shown in Fig. 22.7 restoring the correct UCS setting in each case.
2. To dimension the auxiliary, restore the UCS SLOPE in the active viewport – I must confess that my Fig. 22.7 has a bit of a cheat in it. This is with the leader dimensions for the holes and the true shape 'half width' distance. If you dimension as shown, you should discover why I had to do this!
3. Figure 22.7 has been plotted with the VPBORDER layer frozen. This gives the completed drawing a more realistic effect.

❏ *Summary*

This has not been an easy exercise to complete. If you have reached this stage I hope that you have followed the steps and understood what has been done in each of them. The main points in the exercise are the SOLFEAT, SOLSECT and SOLPROF extractions. This is the method of obtaining extra views from a composite solid and how a traditional 2D type drawing is produced from a real 3D object. The next chapter expands further on this theme.

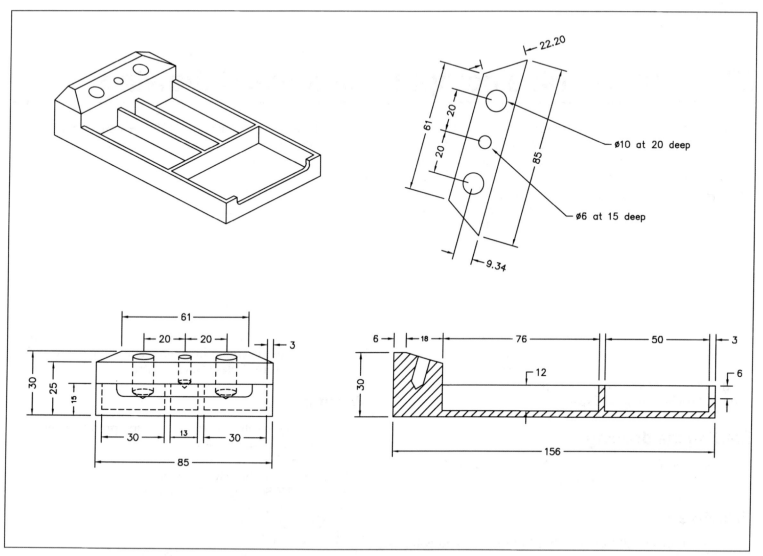

Fig. 22.7. Desk tidy – dimensioned.

23. Exercise 14: a computer link detail drawing

This exercise will create a composite from five primitives, and then extract two profiles to represent an auxiliary view and a true shape. It is a fairly complicated exercise in that it is necessary to create different UCSs as the component is built up. The complete drawing is shown in Fig. 23.1 to give the layout and give the sizes. The steps in the process are fully given, but most of the prompts have been omitted.

Note

1. When entering data try and reason out my figures from Fig. 23.1.
2. It will be necessary to zoom in on areas of the viewports, and I would recommend that you do this using paper space, i.e.
   (a) with the working viewport active, enter paper space
   (b) zoom in on the required area for working
   (c) return to model space and complete work
   (d) enter paper space and zoom previous
   (e) return back to model space.

Starting the drawing

1. Begin a new drawing \SOLID\LINK=\SOLID\SOLA3.
2. Check SOLIDS is the current layer.

Primitive 1

The first primitive will form the 'base' of the link and will be constructed from an extruded vertical polyline.

1. With the upper right viewport active, restore UCS LEFT.
2. Use PLINE to draw a polyline
   from 0,50
   to @50,0
   to @0,−40
   arc to @−10,−10
   line to @−40,0
   arc to @−10,10
   line to close
3. The view is not centred, so ZOOM C about 30,50,30 at 1XP in all viewports.
4. Extrude the polyline for a height of 3 at 0 taper.
5. Create a cylinder at 30,25,0 with diameter 12 and height 3.
6. Create another cylinder at 30,40,0, diameter 6 and height 3.
7. Polar array this cylinder three times about the point 30,25.
8. Subtract the three holes from the extrusion.

Primitive 2

This is a wedge created on top of the first primitive and then rotated into a horizontal position.

1. Keep UCS LEFT.
2. Make lower left the active viewport.
3. Set a new UCS using the 3 point option with:
   (a) origin at 0,50,0
   (b) X axis at 60,50,0
   (c) Y axis at 0,50,3
4. Check UCS at origin – UCSICON OR?

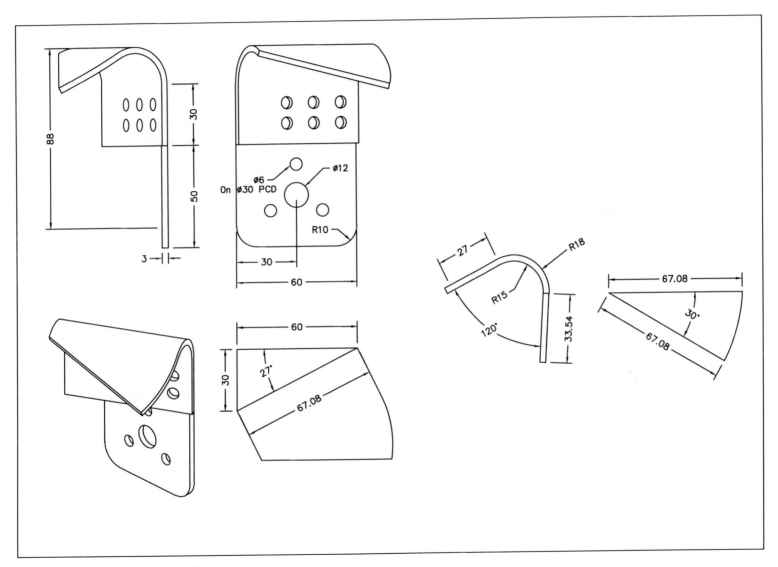

Fig. 23.1. Dimensioned computer link.

5. Save this UCS as PRIM2.
6. Create a wedge (SOLWEDGE) at 0,0,0 with length 60, width −3 and height −30. It will be in the wrong position.
7. Change the colour of this wedge to green – CHPROP
8. Rotate the existing UCS about the Y axis by 90°.
9. Rotate the green wedge by 90° about the point 0,0.

Primitive 3

This primitive is a box created on the edge of the wedge.
1. Restore UCS PRIM2 –probably still in it?
2. Set a new UCS with the 3 point option with:
   (a) origin at 60,0,0
   (b) X axis at 0,30,0
   (c) Y axis at 60,0,-3
3. Save the UCS as PRIM3.
4. Create a box at 0,0,0 with length 67.08, width 30 and height −3. Any idea where the 67.08 size comes from? – think triangles.
5. Change the colour of this box to blue with CHPROP.
6. Create a cylinder at 10,10,0 with diameter 6 and height −3.
7. Rectangular array this cylinder by two rows and three columns, the row distance being 10 and the column distance 15.
8. Subtract the six holes from the blue box.

Primitive 4

This is a polyline which is revolved about an entity.
1. Restore UCS PRIM3 if needed.
2. Set a new UCS on the top edge of the blue box using the 3 point option by entering:
   (a) origin 0,30,−3
   (b) X axis 67.08,30,−3
   (c) Y axis 0,30,0

Note: you could zoom in on the edge and use OSNAP END to pick the points.
3. Save this UCS as PRIM4.
4. Zoom in on the 'free edge' of the blue box.
5. Draw construction lines from 0,0 to @0,−15 to @50,0.
6. Use PLINE to draw a rectangular polyline around the top 'rim' the blue box. Use OSNAP END for this.
7. Change this polyline to yellow (use LAST option).
8. Now SOLREV and pick the yellow polyline (last). Use the Entity option and pick the left end of the long red line as the entity axis. The angle is to be 120°.
   Note: if the revolution goes the 'wrong way', undo and repeat the SOLREV command, picking the other end of the line as the entity.
9. Erase the two red construction lines.
10. Zoom previous.

Primitive 5

The last primitive is another revolution.
1. UCS PRIM4?
2. Set a new UCS to the lower free edge of the yellow primitive with OSNAP END and:
   (a) origin at lower left hand corner of yellow edge
   (b) X axis along bottom lower edge
   (c) Y axis along short (3) edge.
3. Save this UCS as PRIM5.
4. Draw a polyline rectangle around the free end of the yellow primitive and change its colour to cyan.
5. Rotate the UCS about the X axis by 180°.
6. SOLREV the cyan polyline about the Y axis by 30°.

Final composite

1. Restore WCS.
2. Union all five primitives to give the composite computer link.
3. Result as Fig. 23.2.
4. Save your drawing at this stage – SOLMESH,HIDE,SHADE?

New viewports

To extract additional views, new viewports are needed. The handles of these viewports will be unknown and your handle numbers will be different from mine. I'll refer to the two new viewports as VP1 and VP2 as shown in Fig. 23.3.

1. Lower left viewport still active?
2. Enter paper space.
3. Use the MOVE and STRETCH edit commands to 'slim down' the four existing viewports.
4. Use MVIEW to create two new viewports in a position similar to those in Fig. 23.3.
5. Use the LIST command and select each viewport border to obtain the new handle numbers. Note these by sketching on paper your 6 viewports and entering the handle numbers. My two new handle numbers were: VP1 – 6D9
VP2 – 6DA.

Extracting two new views

1. Return to model space with SOLIDS the current layer.
2. Make viewport VP1 (6D9) active.
3. Restore UCS PRIM5 and rotate it about the Y axis by –90°.
4. Enter PLAN to current UCS. This will give an auxiliary view on the end of the link, i.e. the cyan, yellow and blue primitives are viewed 'end on'.
5. This view is out of orientation so enter **DVIEW** and use the **TW** option, entering a twist value of –63 degrees (any ideas why this figure is being used?)

6. ZOOM 1XP (**NOT ZOOM C**) for full size.
7. In VP2 (6DA) restore UCS PRIM5 and rotate about the X axis by -90
8. PLAN to current UCS to give a true shape?
9. ZOOM 1XP to give Fig. 23.4.

New layers

Layers have to be created for the two new viewports to enable 2D profile extraction and dimensioning of the additional views.

1. From the screen menu select **MVIEW**
          VPLAYER
          Newfrz

AutoCAD prompts	New viewport frozen layer names
enter	**DIM-6D9,PV-6D9,PH-6D9<R>** – VP1 handles
AutoCAD prompts	?/Freeze..................
respond	**select Newfrz**
AutoCAD prompts	New viewport frozen layer names
enter	**DIM-6DA,PV-6DA,PH-6DA<R>** – VP2 handles

2. Use the layer control dialogue box to:
   (a) change new DIM layer colour to magenta
   (b) change new PV layer colour to green
   (c) change new PH layer colour to colour number 9
   (d) change the linetype of the new DIM layers to HIDDEN.

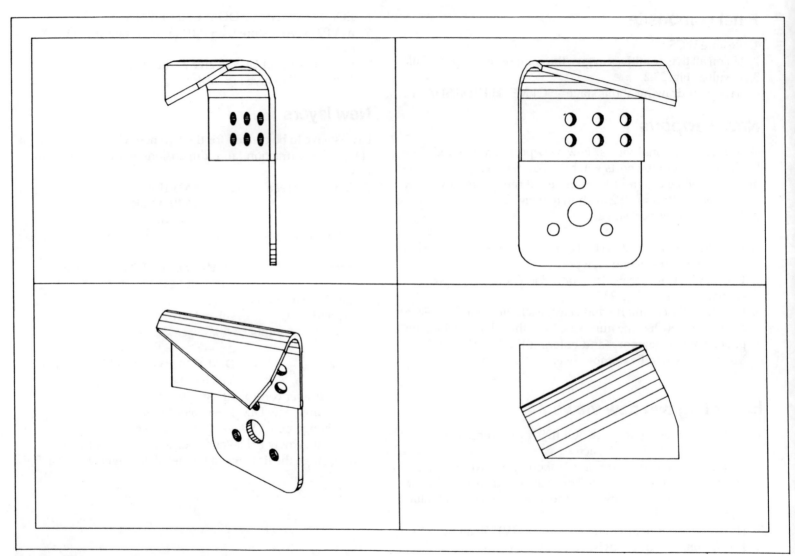

Fig. 23.2. Computer link – completed primitives plottted with HIDE.

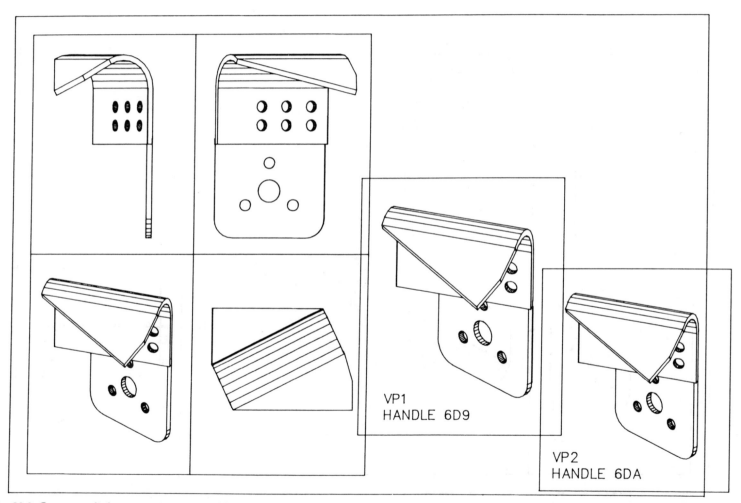

Fig. 23.3. Computer link – new viewports added plotted with HIDE.

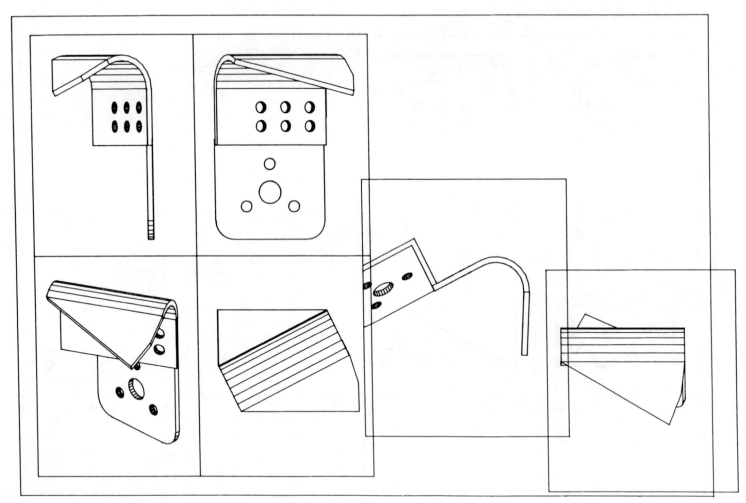

Fig. 23.4. Computer link – viewports with extra views with HIDE.

Extracting the features for the auxiliary and the true shape

1. In model space with viewport VP2 (handle 6DA) active, use the layer control dialogue box to make PV-6DA current. Check that the other PV layers are frozen.
2. Enter **SOLFEAT<R>** and in response to the prompts:
   (a) enter **F<R>** for face
   (b) enter **S<R>** for select
   (c) **pick the cyan shape<R><R>**.
3. AutoCAD should draw in a green–black dotted effect around the cyan outline.
4. In the **other five viewports**:
   (a) make each one active in turn
   (b) make PV-?? the current layer for that viewport – this is not essential!
   (c) enter **SOLPROF<R>** and enter **Y** to the three prompts.
   (d) AutoCAD will draw in a green outline and a rust-coloured hidden line effect over the composite
   (e) repeat (a)–(c) in all viewports.
5. Freeze layer SOLIDS to leave only green continuous outlines and rust hidden lines – I hope!
6. This is shown in Fig. 23.5.

Erasing unwanted detail

1. In model space use the ERASE command and pick each hidden line block in the four 'main' viewports.
2. In VP2 erase the hidden detail then ZOOM 1XP to obtain a full size true shape.
3. In VP1, erase the hidden detail, then EXPLODE the green block – it should turn black.
4. Now erase all unwanted detail from this viewport, to leave only the 'end view' entities.
5. Change the block colour back to green using CHPROP and the bylayer option – this is not essential.

6. Still in VP1 ZOOM 1XP for a full size auxiliary.
7. All being well your drawing should resemble Fig. 23.6.

Viewport alignment

Lining up the viewports is easier than you would imagine.
1. Enter paper space and using the MOVE command, pick the yellow outline of VP1, then

   | at base prompt | pick upper left corner of the green shape |
   | at second prompt | pick the point corresponding to this in the bottom right viewport – top right where three lines meet? |

2. Now MOVE this viewport from any base point (0,0) by **@50<27**.
   Any reason for the angle of 27°?
3. Using MOVE select the yellow border of VP2 and move it in a similar manner. I picked the following: base point: left intersection of the two lines second point: top of right vertical line in VP1.
4. Now MOVE this viewport from a base point (0,0) by **@30<0**
5. These moves should result in Fig. 23.7.

Dimensioning the drawing

This should not give any problems, if you remember to:

(a) make the DIM-? layer current in the active viewport
(b) check all other DIM layers are frozen
(c) restore the appropriate UCS
(d) move to each viewport in turn.

The complete dimensioned drawing is Fig. 23.1 – last is first!

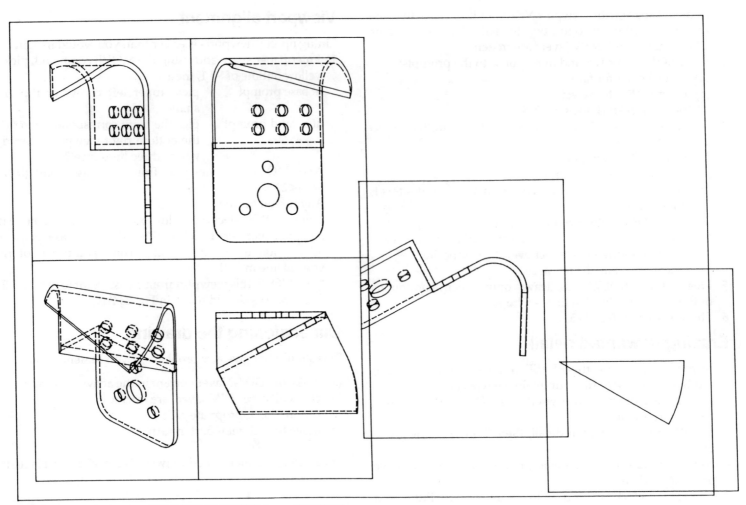

Fig. 23.5. Computer link– after SOLFEAT, SOLPROF: layer SOLIDS frozen.

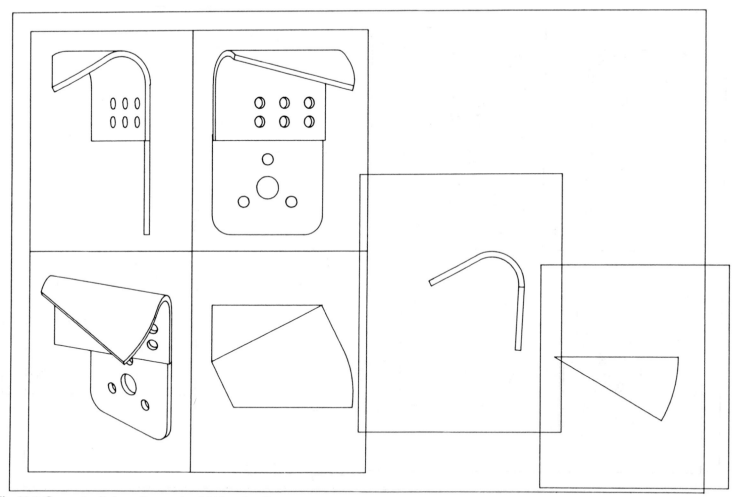

Fig. 23.6. Computer link – after erasing hidden detail.

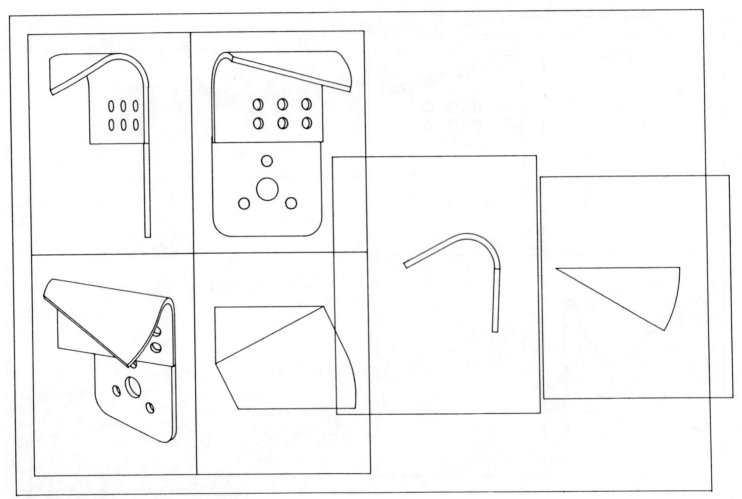

Fig. 23.7. Computer link – after viewport alignment.

24. Exercise 15: block assembly

This exercise will introduce solid model blocks and Xrefs as an assembly drawing. The component is a desk filing system made of two trays and four support legs and Fig. 24.1 gives the details and sizes of the two components.

If you intend plotting any of the drawings in this exercise then I suggest that you alter the SOLWDENS value to 3 or 4 otherwise your paper may end up with holes in it. This happened to me with a SOLWDENS value of 6.

The tray

1. Begin a new drawing **\SOLID\TRAY=\SOLID\SOLA3**.
2. ZOOM C about 55,40,20 at 0.75XP in all viewports.
3. In model space make the lower left viewport active, and make a new layer TRAY, colour green and current.
4. Set SOLWDENS to 4?
5. Create a box at 0,0,0 with length 110, width 80 and height 40.
6. Create a cylinder at 10,0,0 with diameter 20 and height 40.
7. Rectangular array this cylinder for two rows and two columns with the row distance 80 and the column distance 70.
8. Union the four cylinders with the box.
9. Create a box at 10,10,10 – length 110, width 60, height 40.
10. Create a wedge at 110,0,40 – length –15, width 80, height –40.
11. Subtract the box and wedge from the composite.
12. Create a cylinder at 10,0,0 with diameter 10 and height 10.
13. From the menu bar select **Construct**
                  Array 3D
    Pick the cylinder with

two rows, distance 80
two columns, distance 70
two levels, distance 30.

14. Subtract the eight 'holes' from the composite – zoom?
15. Save the drawing as **\SOLID\TRAY** – Fig. 24.2.

The support

1. Begin a new drawing **\SOLID\SUPPORT=\SOLID\SOLA3**.
2. In model space with lower left viewport active make a new layer SUPPORT colour yellow and current.
3. ZOOM C about 0,0,60 at 0.75XP in all viewports.
4. SOLWDENS 4?
5. Create a cylinder at 0,0,0 with diameter 10 and **height 130**. The height of 130 is different from the drawing size of 120, but there is a reason for this – it is not a mistake.
6. Save as **\SOLID\SUPPORT** – Fig. 24.3.

Insert 1

1. Begin a new drawing **\SOLID\ASSEMBLY=\SOLID\SOLA3**.
2. SOLWDENS 4 if you are plotting.
3. ZOOM C about 55,40,90 at 0.5XP in all viewports.
4. Make two new layers, TRAY colour green, SUPPORT colour yellow.
5. Make the lower left viewport active and layer TRAY current.

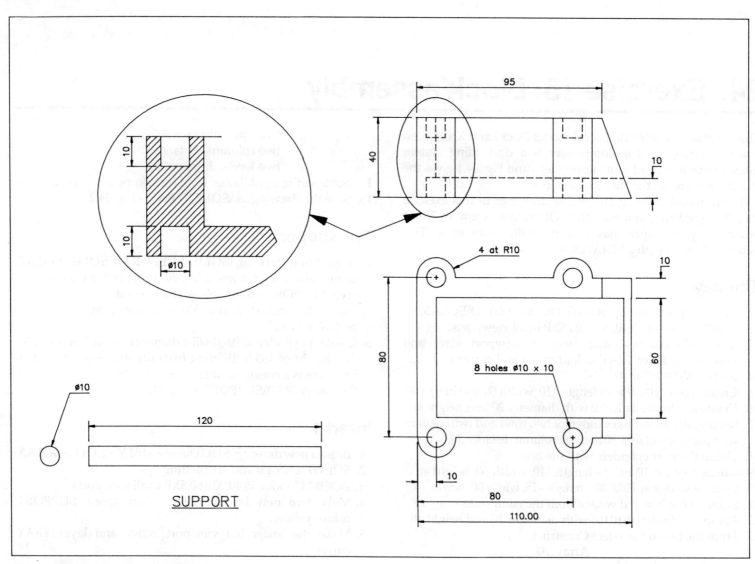

95

40

10

10

ø10

10

10

120

SUPPORT

4 at R10

10

80

60

8 holes ø10 x 10

10

80

110.00

Fig. 24.1. Tray and support details.

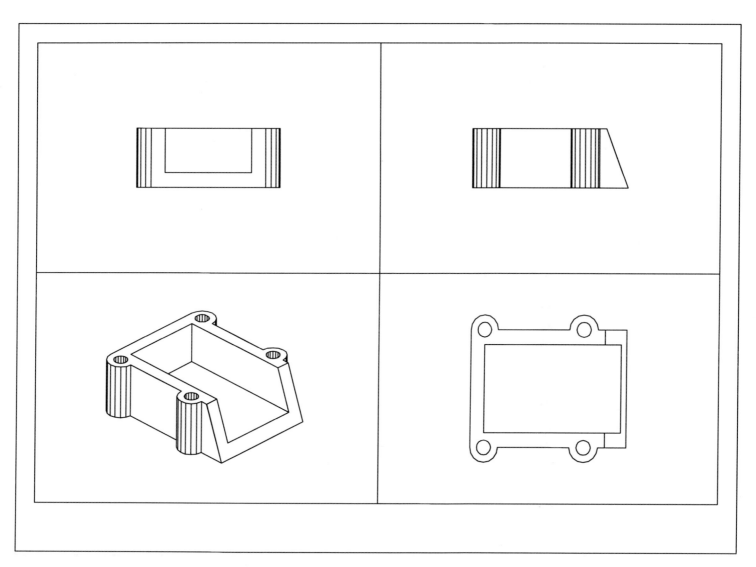

Fig. 24.2. Tray as a composite plotted with HIDE.

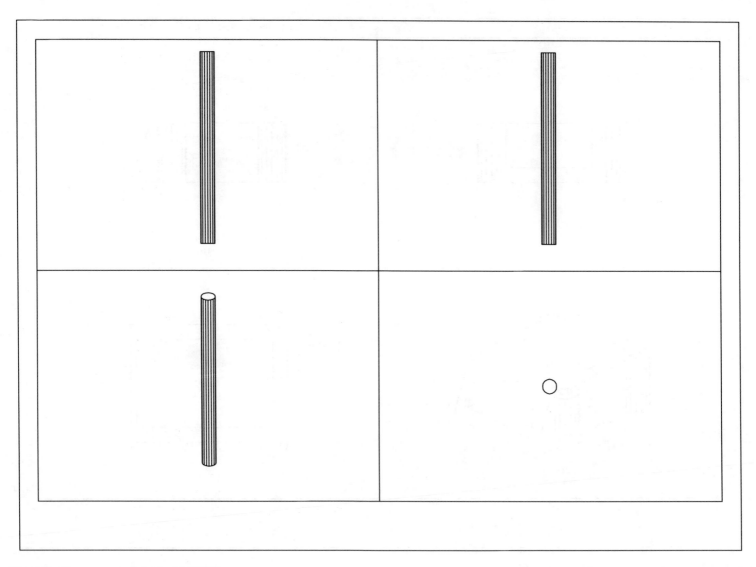

Fig. 24.3. The support plotted with HIDE.

6. From the screen menu select **BLOCKS**
                          INSERT
                          File...
AutoCAD prompts with the Select Drawing File dialogue box
respond   **pick \ <R>**
pick SOLID<R>   – directory name
pick TRAY   – drawing name
pick OK
Enter 0,0,0 as the insertion point and accept the three defaults for the scale and rotation.

7. Enter **INSERT<R>**.
AutoCAD prompts `Block name<TRAY>`
enter   **<RETURN>**, i.e. accept TRAY
AutoCAD prompts `Insertion point`
enter   **0,0,140<R>** and accept the three defaults.
Can you work out the insertion point coordinates?

8. Make SUPPORT the current layer.

9. From the menu bar select **File**
                          Xref>
                          Attach...
AutoCAD prompts with the Select File to Attach dialogue box
respond   check \SOLID is directory name
then   **pick SUPPORT**
                 pick OK
AutoCAD prompts `Insertion point`
enter   **10,0,30<R>** and accept three defaults.
The yellow support will be inserted but will 'interfere' with the top tray due to its length being 130 instead of 120.

10. Important: **save at this stage as \SOLID\ASSEMBLY** – Fig. 24.4.

There are several points worth mentioning:
(a) the prompt with SAVE – Scanning...
(b) the interference of the yellow support with the top tray
(c) the layer control dialogue box displays layers of the format **SUPPORT |.................**

11. Use SOLLIST and pick the tray.
AutoCAD prompts `1 object ignored`
                 `No objects found`
Why is this? At this stage the tray and support inserts are not yet solids. They must be bound and exploded before AutoCAD will recognise them as solids.

12. At the command prompt, enter **XREF<R>**
AutoCAD prompts `?/Bind/Detach............`
enter   **B<R>**
AutoCAD prompts `Xref(s) to bind`
enter   **SUPPORT<R>**
AutoCAD prompts `Scanning...`
then   returns the prompt line.

13. Check the Layer Control dialogue box. The SUPPORT |.... layers have been replaced by **SUPPORT0........** layers.

14. EXPLODE the two green trays and the yellow support.

15. Now use SOLLIST on the tray object – SUBTRACTION, etc. and SOLLIST on the support object – CYLINDER, etc.

16. From the menu bar select **Model**
                          Inquiry
                          Interference
AutoCAD prompts `Select the first set of`
                 `solids...`
                 `Select objects`
respond   **pick the top green tray<R>**
AutoCAD prompts `1 solid selected`
                 `Select the second set of solids...`
                 `Select objects`
respond   **pick the yellow support<R>**

AutoCAD prompts   `1 solid selected`
                          `Comparing 1 solid against 1 solid`
                          `Interfering solids`
                                `(first set):     1`
                                `(second set):    1`
                          `Interfering pairs     1`
                          `Create interference solids <N>`
enter   **Y<R>**
AutoCAD prompts   by displaying a yellow cylinder in the top two viewports. This is the interference solid between the yellow support and the top tray.

17. Proceed immediately to the next section.

Insert 2

1. Open the **\SOLID\SUPPORT** drawing saved earlier and discard the changes, i.e. do not save the interference steps.
2. Use SOLCHP to alter the length of the support to its correct value of 120.
3. SAVE as **\SOLID\SUPPORT**.
4. Open the **drawing \SOLID\ASSEMBLY** from step 10 of the Insert 1 section (on page 157) – hope you did save it! The yellow support should be displayed without any interference, i.e. it should fit into the top tray. This is because it is a bound Xref in the ASSEMBLY drawing, and the alterations to the SUPPORT drawing automatically altered the sizes in the ASSEMBLY drawing.

   During the opening process there are various prompts which may be new to the reader, e.g.
         `Resolve Xref SUPPORT: SUPPORT.dwg`
         `Duplicate registered.............`
         `SUPPORT loaded`
5. Rectangular array the yellow support for two rows and two columns, the row distance being 80 and the column distance 70.

6. Use SOLLIST to check if any of the objects are solids?
7. Convert the imported blocks by binding the Xref as step 12 in the Insert 1 section.
8. Explode all six blocks and SOLLIST to check they are solids.
9. Union all six solids to give the completed assembly – Fig. 24.5.
10. Save the completed assembly as **\SOLID\COMASSEM.**

'Cutting' the assembly

1. Using the complete assembly drawing, create a box at 42,40,0 with length 90, width 90 and height 200.
2. Change the colour of this box to blue with CHPROP.
3. Rotate this blue box about 42,40,0 by -45°.
4. Subtract the box from the composite – Fig. 24.6.
5. Try SOLMESH, HIDE and SHADE then return to wire-frame.

Sectioning the assembly

1. Open the complete assembly drawing **\SOLID\COMASSEM** and ensure the lower left viewport is active.
2. In paper space erase the top two viewports and the bottom right viewport.
3. Make VPBORDER the current layer.
4. From the menu bar select **View**
                         MView
                         3Viewports
                         enter **V<R>** for vertical
                         then **10,25<R>** for first point
                         370,260<R> for second point
5. Erase the 'small' viewport and redraw.
6. In model space ZOOM C about 55,40,90 at 0.75XP in the three viewports.
7. Use VPLAYER with the Newfrz option to make three new layers – SECTLEFT, SECTMID, SECTRIGHT. The colour is to be magenta.

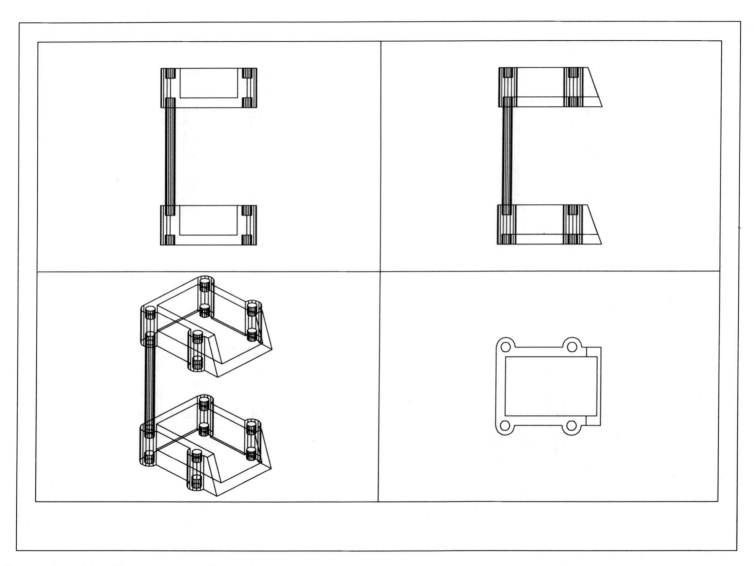

Fig. 24.4. Assembly with support inserted as Xref.

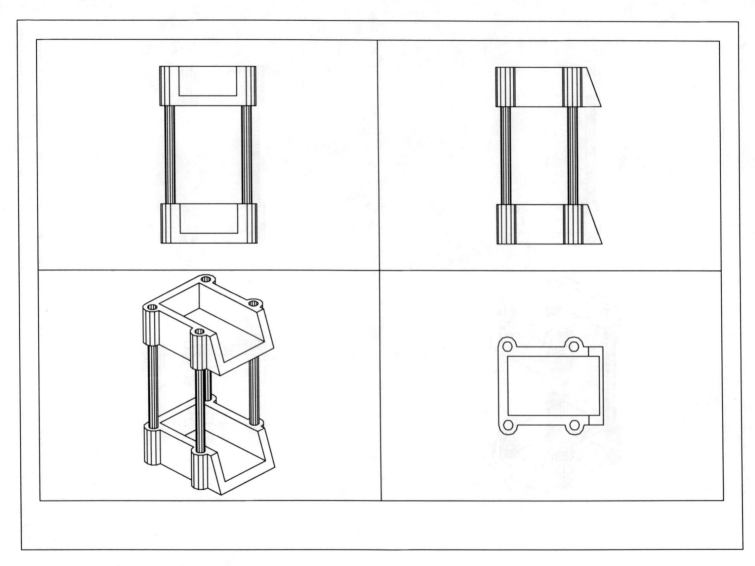

Fig. 24.5. Completed assembly plotted with HIDE.

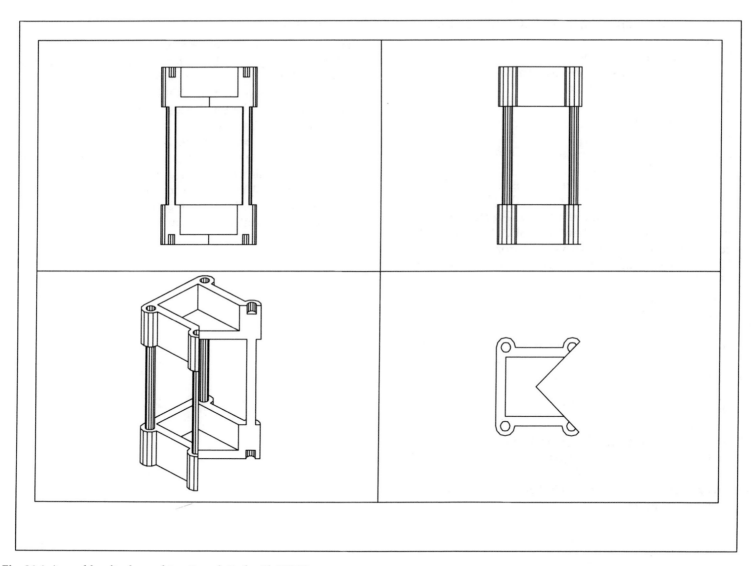

Fig. 24.6. Assembly after box subtraction plotted with HIDE.

8. In the left viewport, thaw layer SECTLEFT and make it current.
9. UCSICON All and ORigin.
10. Create a new UCS using the 3 point option with:
    (a) origin at 10,0,0
    (b) X axis at 80,80,0
    (c) Y axis at 10,0,40.
11. Save this UCS as SECT1.
12. Set the following variables – SOLHPAT to U, SOLHSIZE to 3 and SOLHANGLE to 45.
13. Enter **SOLSECT** and pick the composite. Enter *XY* as the section plane with 0,0,0 as a point on the plane.
14. A magenta section should be added as Fig. 24.7(a).
15. In the middle viewport, thaw layer SECTMID and make it current. Check other SECT layers are frozen.
16. Rotate the UCS about the *X* axis by –90 and save it as SECT2.
17. SOLSECT and pick the composite. Enter *XY* as the plane and 0,0,50 as a point on the plane.
18. The section should be as Fig. 24.7(b).
19. In the right viewport restore WCS, thaw layer SECTRIGHT and make it current. Check other SECT layers are frozen.
20. Make a new UCS using 3 point
    (a) origin at 80,0,0
    (b) *X* axis at 80,80,0
    (c) *Y* axis at 0,0,0.
21. Rotate the UCS about *X* by 63° and save it as SECT3.
22. SOLSECT, pick the composite, enter *XY* and 0,0,0 – Fig. 24.7(c).
23. Freeze layers TRAY and SUPPORT to leave the three sections as Fig. 24.7.

Points worth mentioning

1. The layers during the xref attachment were:
   SUPPORT   – layers for Xrefs attached to drawings
   SUPPORT0 – layers for bound Xrefs.
2. If a directory is taken for the **SOLID** directory, all the files in that directory will be listed. Most of these files will have the extension .DWG as they are drawing files. There should also be a file with the extension **.XLG** – ASSEMBLY.XLG – which may be new to the reader. This is an Xref log file (XLG) and is a text file. This means that it can be 'viewed' using the TYPE ASSEMBLY.XLG command. The file contains information about the xrefs to be loaded into a drawing. If you have a printer you can list the file with TYPE ASSEMBLY.XLG>PRN.

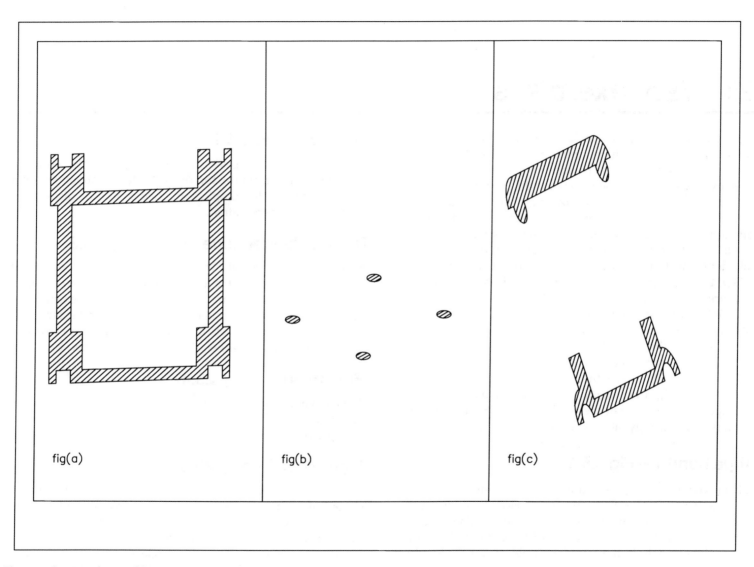

Fig. 24.7. Sectioned assembly.

25. User exercises

Now that the instructions in solid modelling techniques are complete, it is only with continual practice that the user will become proficient. For this reason this chapter is devoted to user exercises. I have drawn several different types of solid models in the accompanying drawings, but have not added any sizes. The drawings are meant to act as a guide, and in attempting them sizes are left to the user's discretion.

All the drawings have been completed on the SOLA3 standard sheet with the four viewport configuration. The ZOOM C effect can be quite tricky to obtain but will be good practice for you. I have added some comments about the drawings, but it is really for you to decide on a strategy for obtaining the final composite. There are a few 'pipe type composites' as these are ideal to demonstrate solid modelling.

When each composite is complete, remember SOLMESH, HIDE and SHADE to 'see' the solid.

Several of the drawings involve quite a bit of computer manipulation, and if you are working on a 386 machine (as I was) the waiting may take some time.

Pipe bend 1 – Fig. 25.1

This involves a few UCS settings. The basic idea is to draw a circle and SOLREV it about an axis. I used the entity option (with a line) which can be awkward as you invariably pick the wrong end of the line and the solid is revolved the 'wrong way'. Colours help in the selection of objects for the SOLREV command. Union all primitives to obtain the composite.

Tail pipe – Fig. 25.2

This idea came for a gas-turbine engine I used to work on. SOLREV a circle from the top of a cylinder for about 60°, then array four times. The complete unioned composite displays an interesting interpenetration.

Double flanged pipe bend – Fig. 25.3

A circle SOLREV for 180°. The flanges are cylinders of different diameters and heights. One contains holes, the other an extruded polyline and cylinders. Array is used a few times. Remember to add an 'inside pipe' to simulate wall thickness and subtract the inside pipe for the composite. The flanges need not have anything extra added. UCS settings help. Subtraction and union operations.

Pipe bend 2 – Fig. 25.4

A relatively simple pipe bend. Copy may make things easier than you think. Two unioned cylinders are added as an afterthought, but union all primitives.

Pipe bend 3 – Fig. 25.5

I admit that I went a bit 'over the top' with this drawing. It started off as a figure 8, but somewhere along the line I lost track of what I was doing. I had to add straight cylinders to finish off the composite, and the inclined part (using rotation) caused some concern. Basically a circle is SOLREVed about an axis, the problem being to define the UCS properly.

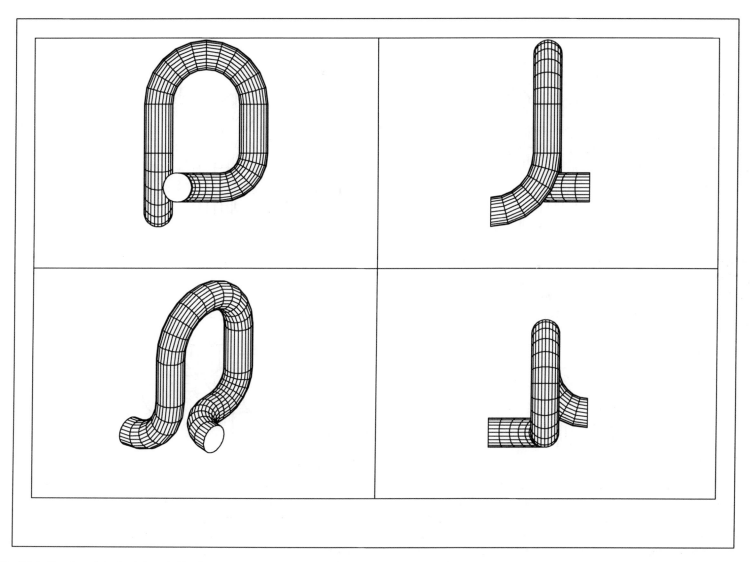

Fig. 25.1. Pipe bend 1 plotted with HIDE.

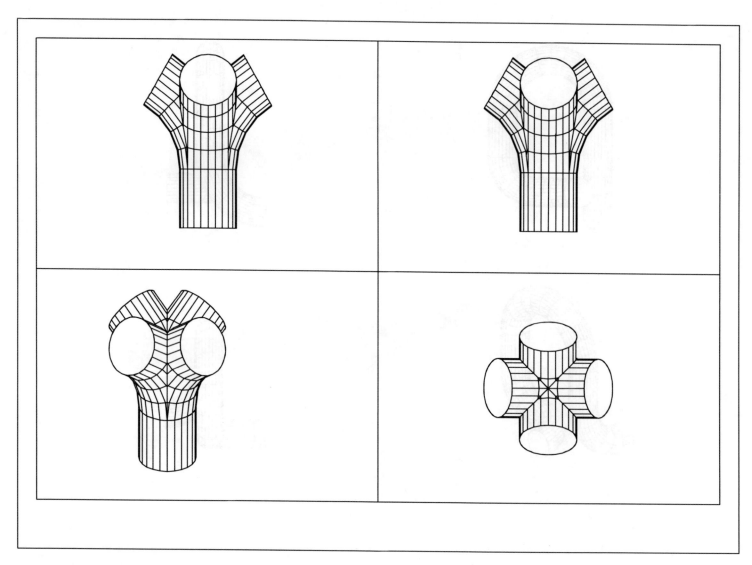

Fig. 25.2. Tail pipe exhaust plotted with HIDE.

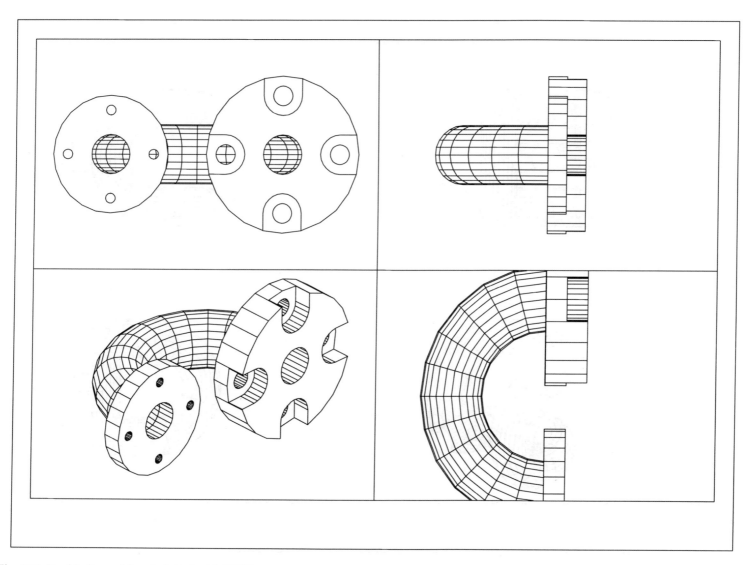

Fig. 25.3. Double flanged bend plotted with HIDE.

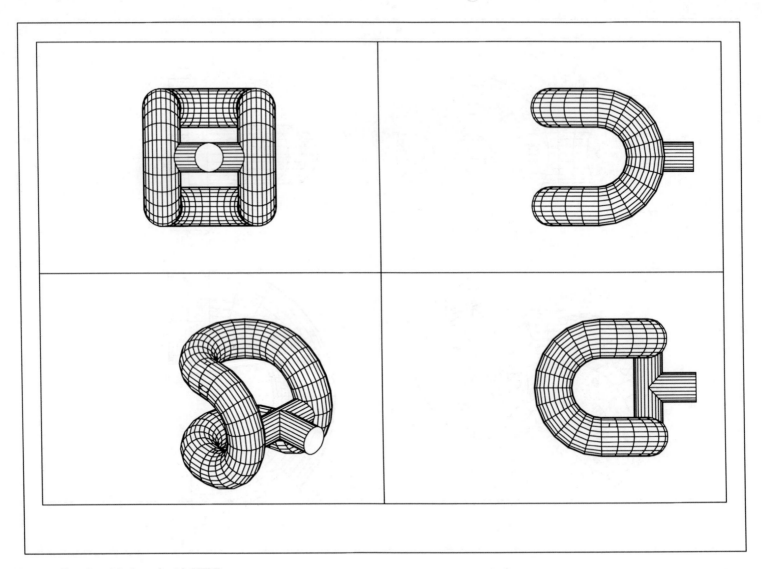

Fig. 25.4. Pipe bend 2 plotted with HIDE.

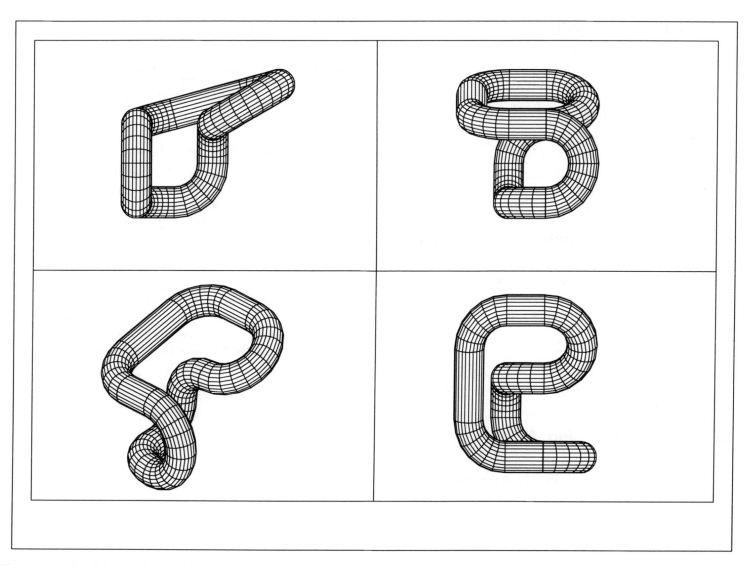

Fig. 25.5 Pipe bend 3 plotted with HIDE.

Greek temple – Fig. 25.6

A simple solid consisting of boxes and arrayed cylinders. The top is an extruded polyline. This drawing could have been obtained using standard AutoCAD 3D techniques.

A casting – Fig. 25.7

Contains more than I had intended. The basic casting is a cube, with another cube subtracted from it. The bottom of the inside cube is from a sphere using subtraction. The bottom uses SOLFILL and the top SOLCHAM. the slot and hole effect are extras. Subtraction is used a lot.

Cone/cylinder interpenetration – Fig. 25.8

Traditional type of problem for 2D geometry which is simple with solid modelling. Of the two cylinders, the inclined one may cause you a few problems. All three primitives are unioned to obtain the interpenetration effect.

Elliptical pipe – Fig. 25.9

One cylinder subtracted form another, then SOLCHP used to make the elliptical pipe. The holes are from the subtraction of a box and an arrayed cylinder.

A thin-walled shell – Fig. 25.10

Two sets of spheres unioned with a cylinder – outside and inside. I used radii of 50 and 45. The holes are subtracted cylinders, and SOLSECT was used to obtain the section detail.

SOLBOX structure – Fig. 25.11

A simple exercise using only SOLBOX. UCS settings are necessary and the basic box shape is the same every time. I used a box that was $80 \times 30 \times 20$. Changing colour helps when shading is added. The end result is dependent on your imagination.

Newton's cradle – Fig. 25.12

A nice practical application of solid modelling. The cradle is made from cylinders and by revolving a circle about an entity. I used array for the legs and corners, then rotated the corners into position. Spheres and cylinders are obvious for the balls, but the wire holding them to the cradle is interesting. Multiple copy was used, and UCSs are needed.

Can you rearrange the position of the first sphere to be at the start of a swinging motion, at an angle of about 25°. The supporting wire must be 'rotated' as well.

If you know how to make slides and write script files, an interesting digression is to create a slide show to simulate the balls in motion. I achieved this with five slides with the spheres in different positions, and a script file consisting of about a dozen lines.

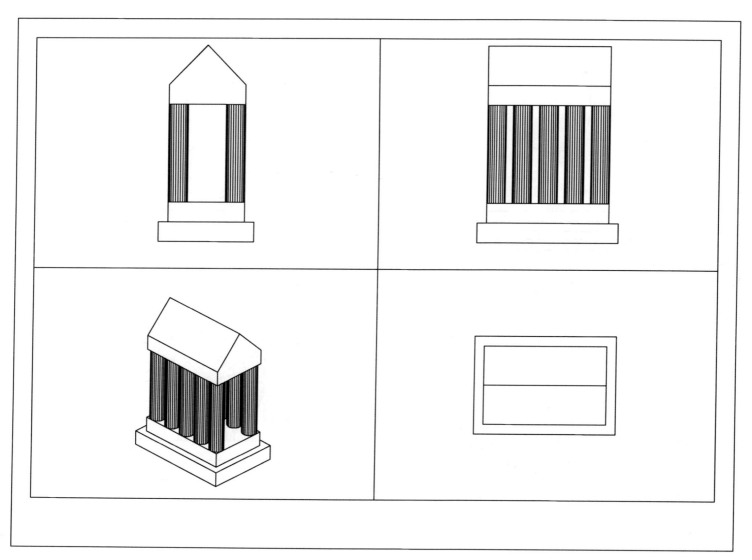

Fig. 25.6. The Greek temple plotted with HIDE.

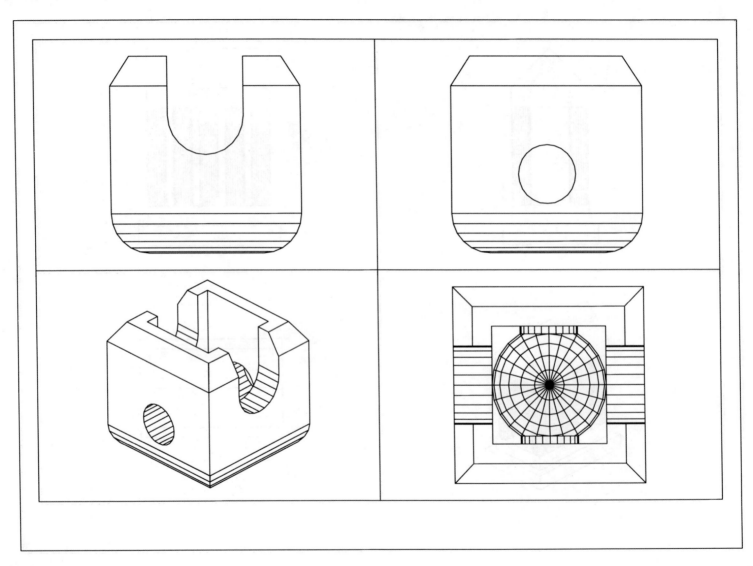

Fig. 25.7. Casting plotted with HIDE.

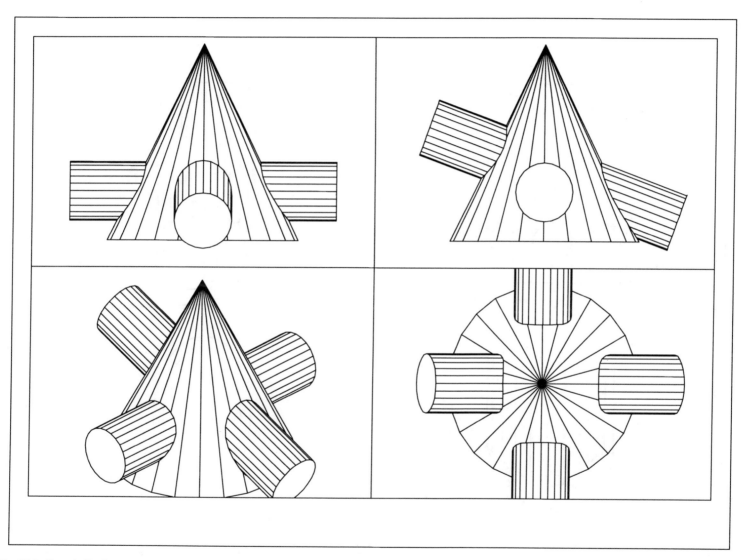

Fig. 25.8. Cone/cylinder interpenetration plotted with HIDE.

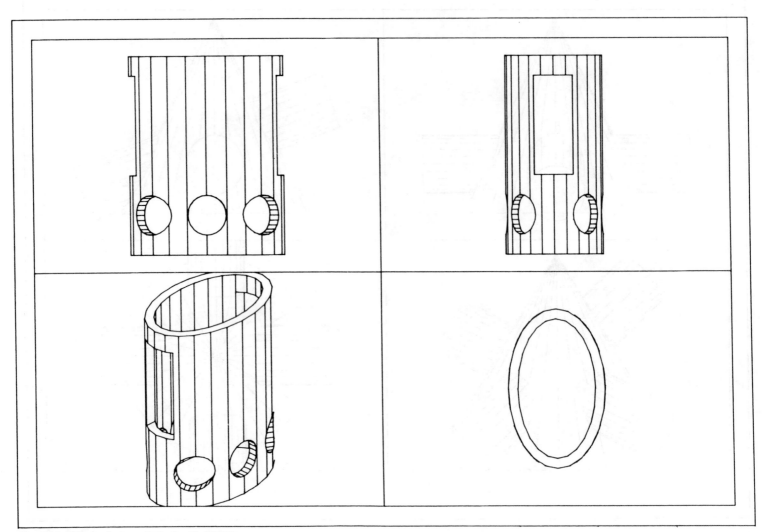

Fig. 25.9. Elliptical pipe plotted with HIDE.

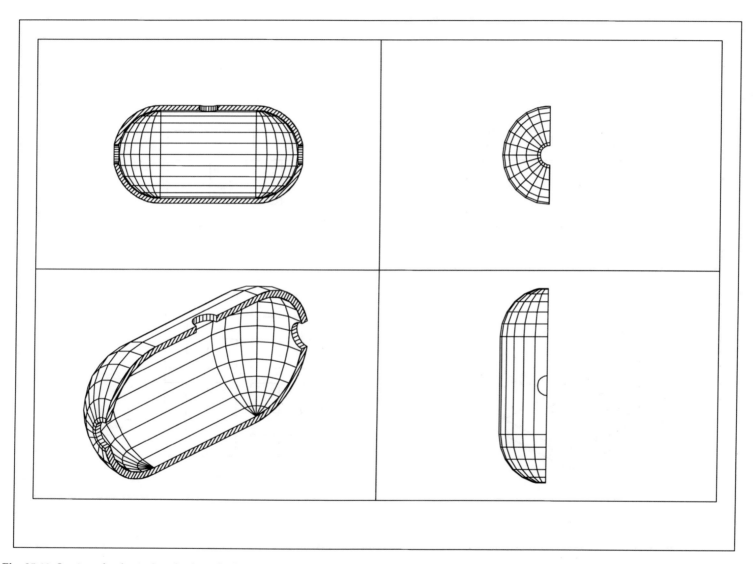

Fig. 25.10. Sectioned spherical tank plotted with HIDE.

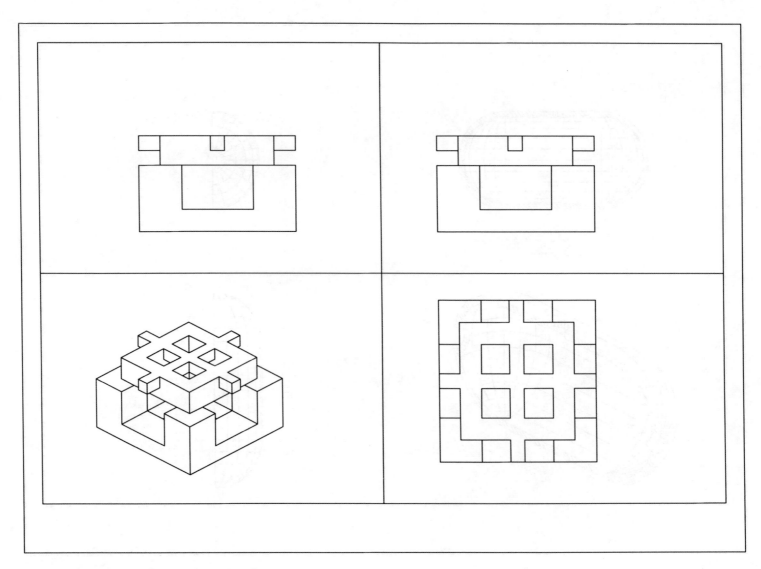

Fig. 25.11. SOLBOX sculpture plotted with HIDE.

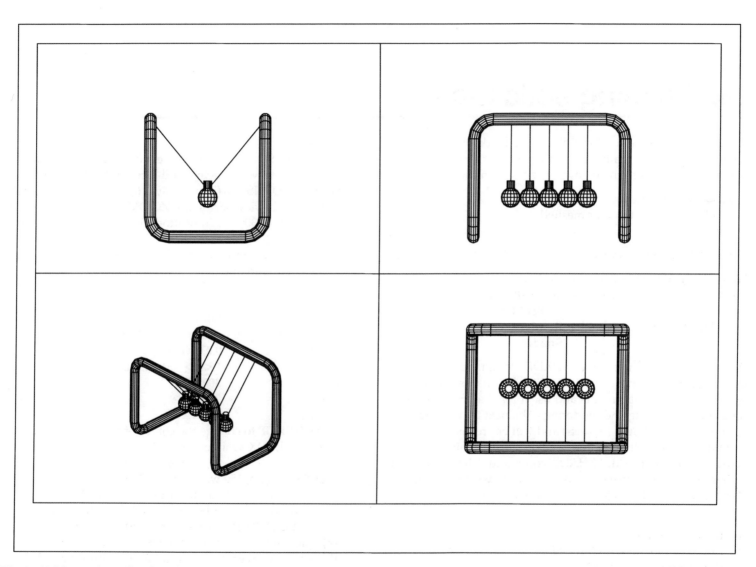

Fig. 25.12. Newton's cradle plotted with HIDE.

26. Shading solid models

The SHADE command allows objects to be displayed in full colour, if their surfaces have been drawn in colour. Shading can only be used with the following types of entities:

- extruded 3D drawings
- 3D objects, either faced or meshed
- solid models.

The actual command is very easy to use and can be activated by

(a) entering **SHADE<R>** at the command line
(b) selecting from the screen menu **DISPLAY**
                                    SHADE
(c) selecting from the menu bar **Render**
                                    Shade

We have already used the SHADE command in several exercises, but have not investigated it fully. In this chapter we will create a new solid model and use it for shading, as well as for dynamic viewing and rendering – the next two chapters.

  The model is a house (of sorts) and will be created from solid primitives – basically box and wedge. Each primitive will be coloured to enhance the SHADE command. The example is rather long, but persevere, as it is well worth the effort. Remember that the solid model commands can be activated from

(a) the command line
(b) the screen menu with MODEL
(c) the menu bar with Model.

The walls

1. Open your SOLA3 standard sheet in model space with the bottom left viewport active and layer SOLIDS current.
2. Set SOLWDENS to 6 (it probably is 6) which will load AME.
3. Centre each viewport with ZOOM C, about the point 150,100,120 at 300 magnification (500 magnification in the 3D viewport).
4. Set the UCS icon to the origin in all viewports by entering at the command line

UCSICON<R>
A<R>
OR<R>

5. At the command line enter **SOLBOX<R>** then

corner	**0,0,0<R>**
then	**L<R>** – the length option
length	**300<R>**
width	**200<R>**
height	**150<R>**

6. Use **CHPROP** to change the colour of the box to YELLOW.
7. Using SOLBOX again enter **10,10,0** as the corner point
                    L the length option
   then **280** length, **180** width and **150** height.
8. This second box should be GREEN, so CHPROP.....
9. The second box is to be subtracted from the first box, so enter **SOLSUB<R>** at the command line and:
   (a) pick the yellow box at the first prompt then**<R>**
   (b) pick the green box at the second prompt then**<R>**
10. We now have a solid composite for the walls.

Adding windows and a door

The door and windows will be represented as box primitives in the walls and then subtracted from the walls to leave 'openings'.

1. Bottom left viewport active – it should be.
2. Use the SOLBOX command with the following values:
   (a) corner: 120,0,0   (b) corner: 30,0,50   (c) corner: 210,0,50
       length: 60   length: 60   length: 60
       width: 10   width: 10   width: 10
       height: 120   height: 60   height: 60
3. These three boxes have to be RED, and they should be, as they have been drawn on the SOLIDS layer.
4. A window is to be added to the 'right' side wall and the UCS needs to be re-positioned so select from the menu bar
   Settings
   UCS
   Named UCS...
   Front-Current-OK
5. Use SOLBOX with the following entries
       corner: 50,50,300
       length: 100
       width: 60
       height: –10
6. This box should be red. Can you reason out the entries in this last SOLBOX command?
7. Subtract the four boxes from the wall composite with the SOLSUB command and
   (a) pick the walls first then<R>
   (b) pick the four red boxes then<R>.
8. We now have a solid composite with four walls and four openings.

Adding the eaves

Eaves will be added to the side walls as solid wedges and then unioned with the walls. To assist with the eave construction, we will set a new UCS position.

1. Restore UCS-World.
2. At the command line enter **UCS<R>** then
       O<R> – new origin option
       0,100,150<R> – new origin position
       UCS<R>
       Z<R>
       –90<R> – rotate about z axis option.
       UCS<R>
       S<R>
       EAVE<R> – save UCS position as EAVE.
3. The above sequence will have moved the UCS icon to the top midpoint of the left side wall with the x-axis pointing to the front and the y-axis pointing towards the right wall.
4. Use SOLWEDGE with the following values:
   (a) corner: 0,0,0   (b) corner: 0,0,0
       length: 120   length: –120
       width: 300   width: 300
       height: 50   height: 50
       colour: yellow   colour: yellow
5. Enter **SOLUNION** and pick the two yellow wedges then<R>.
6. Create another two wedges (colour green) with:
   (a) corner: 0,10,0   (b) corner: 0,10,0
       length: 120   length: –120
       width: 280   width: 280
       height: 50   height: 50
7. Union the two green wedges with SOLUNION.
8. The green wedges have now to be subtracted from the yellow wedges, so enter **SOLSUB** and:
   (a) pick the yellow wedges for the first prompt then<R>
   (b) pick the green wedges for the second prompt then<R>.

9. Enter **SOLUNION** and pick the walls and wedges then **<R>**.
10. The house composite now consists of four walls with a door and three window openings, and eaves added to the side walls. Your drawing should be as Fig. 26.1.
11. At this stage save your model as \SOLID\HOUSE.
12. At the command line enter **SOLMESH<R>** and pick the composite in response to the prompt then **<R>**.
13. From the menu bar select **Render-Shade** and you will have a shaded house in 3D. The SHADE command can be repeated in the other three viewports, and the display is quite nice?
14. Return the model to wire-frame with REGEN and SOLWIRE.

Adding the roof

The roof will be added with a new UCS position and created on a new layer. This will allow us to 'take the roof off' and 'see into the house'.

1. Using the Layer Control dialogue box, make a new layer called ROOF with colour magenta and current.
2. Restore UCS-World.
3. At the command line, enter **UCS<R>** then
   3<R> – the three point option.
   0,–20,150<R> – the origin point
   300,–20,150<R> – the x-axis position
   0,100,200<R> – the y-axis position
   UCS-S-ROOF – saved UCS position.
4. The UCS icon should be aligned along the left hand eave wall. Can you reason out the co-ordinate entries?
5. With **SOLBOX** enter 0,0,0 as the corner, 300 as the length, 130 as the width and 10 as the height.
6. A magenta box (roof) will have been added to the front of the house. An interesting entry is the 130 width. Where did I get this number?

7. Using SOLBOX again, enter 0,130,0 as the corner, 300 as the length, 130 as the width and 10 as the height.
8. This second part of the roof has been added as an 'extension' to the first part, and will have to be rotated into its correct position. To achieve this, select from the menu bar **Modify-Rotate 3D** and

   AutoCAD prompts Select objects
   respond **pick the second magenta wall then<R>**
   AutoCAD prompts Axis by entity.............
   respond **X<R>**, i.e. rotate about the X axis
   AutoCAD prompts Point on x-axis<0,0,0>
   enter **0,130,0<R>**
   AutoCAD prompts <Rotation angle>/Reference
   enter **–45.24<R>**

9. Hopefully the second roof section will have been rotated into position, but why –45.24?
10. Now SOLUNION the two magenta roof sections.

Note

1. The roof has a V at the apex. We could fill this in, but will leave it as it is.
2. The house now consists of two composites – the walls and the roof. Do not union the roof with the walls as
   (a) this will affect the SHADE command
   (b) we want to be able to take the roof off in later.

Adding a floor and an open door

1. Restore UCS-World and make SOLIDS the current layer.
2. Create a floor using SOLBOX with 10,10,0 as the corner, 280 as the length, 180 as the width and 10 as the height.
3. Change this box colour to cyan using the CHPROP command with the (L)ast option.
4. Create a door with SOLBOX and 120,0,0 as the corner, 60 as the length, 10 as the width and 120 as the height.

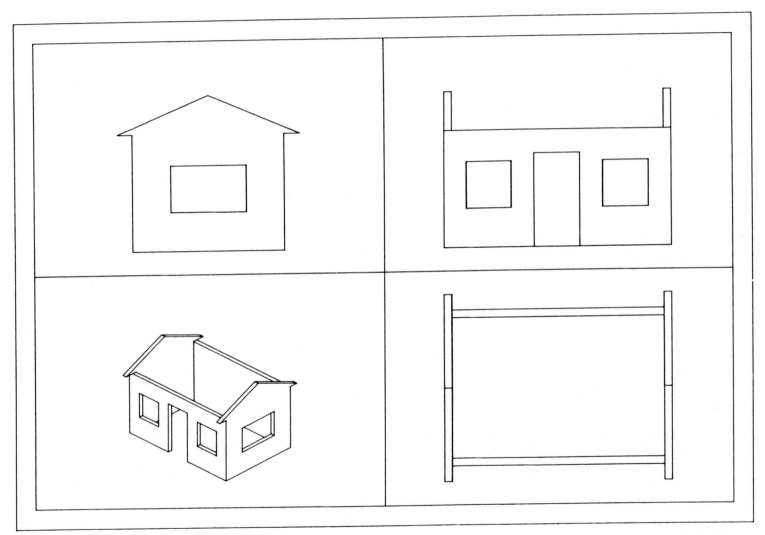

Fig. 26.1. Solid model house after walls, doors, windows, eaves.

5. With CHPROP and the (L)ast option, make this box green.
6. Open the door with **ROTATE** and use the (L)ast option then:
   (a) 120,0 as the base point
   (b) −65 as the rotation angle.
7. Union the floor and open door with the walls.

Adding a chimney

The last item we will add to our house is a chimney.

1. Lower left viewport still active?
2. Restore UCS-World and make ROOF layer current.
3. At the command line enter **SOLCYL<R>** with:

   centre point   50,70,195
   radius   10
   height   30
4. Change the colour of the chimney to blue, and union the magenta roof sections with the blue chimney.
5. You solid model house is now complete – Fig. 26.2.
6. At this stage save your model as **\SOLID\HOUSE** for future work.
7. SOLMESH the complete house then enter SHADE – nice display although the blue chimney is not too clear?
8. REGEN, SOLWIRE to return to wire-frame representation and make layer SOLIDS current.

Shading

When solid models have been created from coloured primitives, the SHADE command gives a nice coloured image of the model. The terminology associated with shading consists of only three words, these being:

- SHADEDGE
- SHADE
- SHADEDIF.

SHADEDGE

1. In all viewports set the viewpoint to VPOINT 'R' 315, 30. Your model house will be displayed in 3D in all viewports.
2. Make the lower left viewport active.
3. Freeze layer ROOF.
4. From the screen menu select **DISPLAY**
                                    SHADE
                                    Shadedge
   AutoCAD prompts   `New value for SHADEDGE<3>`
   enter   **0<R>**
5. Now SOLMESH the walls then enter SHADE**<R>**.
6. Repeat the DISPLAY–SHADE–SHADEDGE sequence then SHADE in each viewport using the following SHADEDGE entries:
   (a) upper left: 1
   (b) upper right: 2
   (c) lower right: 3.
7. The final result will be a coloured display in each viewport, with shading added to different 'degrees'.

SHADEDGE is a system variable which controls the type of rendering produced with the SHADE command. The value of SHADEDGE can be 0-3 and the effect of each value is

SHADEDGE 0: shading is produced without highlighted edges, but requires a 256 colour display.

SHADEDGE 1: creates shading with highlighted edges. These edges have the same colour as the background colour. Also requires a 256 colour display.

SHADEDGE 2: no shading is added, but hidden lines are removed, i.e. it is similar to the HIDE command. This value works with all types of display.

SHADEDGE 3: adds shading by drawing faces with their original colour. Edges are also highlighted in the background colour and works with all types of display.

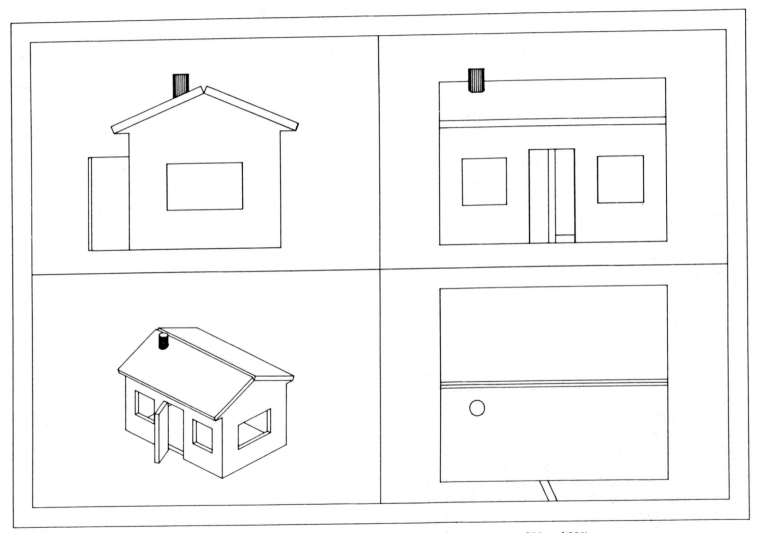

Fig. 26.2. Solid model house with roof, chimney and open door (see also colour plates between pages 200 and 201).

The SHADE screen menu displays four SHADEDGE selections, and these are equivalent to entering 0–3 at the command line as follows:

256-col:	0	256-edg:	1
Hidden :	2	Filled :	3

SHADE

The SHADE command assumes that a light source is available. This light source is said to be 'over-the-shoulder', i.e. it is directly behind the viewpoint irrespective of this viewpoint setting. The actual light source is defaulted such that 70% of the light is being automatically reflected from the source and 30% is taken as ambient background light.

SHADEDIF

1. In all viewports, set SHADEDGE to 1 and make the lower left viewport active.
2. From the screen menu select **DISPLAY**
   SHADE
   Shadedif
   AutoCAD prompts `New value for SHADEDIF<70>`
   enter   **5<R>**
3. Now enter SHADE.
4. In the other three viewports, use DISPLAY–SHADE–Shadedif then SHADE with the following values for Shadedif:
   (a) upper left: 35
   (b) upper right: 65
   (c) lower right: 95.
5. The walls will be displayed with different shading displays.
   The SHADEDIF system variable can be used to the alter the 70%/30% reflected/background light setting. As stated, the default setting is 70, and the higher values will increase the reflectivity and contrast.

This completes the investigation into shading solid models.

❏ *Summary*

1. SHADE is a command which gives coloured solid model images.
2. The SHADE command is rather 'bland'. Colour is added, but there is no impression of depth, roundness, shadows etc.
3. SHADEDGE is a variable which determines the final definition of the shaded component. The default SHADEDGE is 3.
4. A SHADEDGE value of 2 is similar to HIDE.
5. SHADEDIF determines the proportion of light 'falling' on the model. The default setting is 70, i.e. 70% of the available light falls on the component and 30% is background.

27. DVIEW with solid models

The DVIEW command allows the user to view solid models (and other 3D objects) dynamically – hence the name of dynamic viewing. The command introduces perspective to a component and has several options available for selection. It also has it's own terminology. The basic idea of DVIEW is that the user is looking at the component (the **TARGET**) through a **CAMERA** along a **LINE OF SIGHT**. The user can alter the angle of the camera and target relative to the traditional *XY* plane and the target can also be **TWISTED** relative to the line of sight. The camera **DISTANCE** from the target can be increased or decreased and the user can **POINT** the camera at a specified point on the target. It is possible to **ZOOM** and **PAN** the target, and **HIDE** will display the target with hidden line removal. The **CLIP** option allows parts of the target to be 'cut-away' thus allowing viewing 'inside' the object.

Entries can be made via the keyboard or by interactively using the DVIEW block which AutoCAD provides. An additional bonus with the command is that several options can be used one after the other, e.g. the user can use the camera option, then the twist option, then the clip option, then the hide option etc. **UNDO** then allows unwanted options to be 'undone'. This allows a great amount of flexibility with the command.

The DVIEW command can be activated by three methods:
1. From the command line with **DVIEW<R>**.
2. From the screen menu with **DISPLAY**

   DVIEW

3. From the menu bar with **View**

   Set View

   Dview

It is the user's preference as to what method is used, but I will use the keyboard entry in the exercise, as I find it quicker.

When the DVIEW command is activated, the user is prompted to select the object which is to be viewed dynamically, and is then prompted with the following options:

```
CAmera/TArget/Distance/POints/PAn/Zoom/TWist/
CLip/Hide/Off/Undo/<eXit>
```

The various options are activated by selection from the screen menu, or by entering the CAPITAL LETTERS (e.g. CA, TW, CL, U) at the command line.

The example for the DVIEW demonstration will be the solid model house created in the last chapter. We will also add some extra solids to the house, make some slides and then run a simple slide show.

Getting started

1. Open your **\SOLID\HOUSE** solid model drawing in model space with the lower left viewport active. Restore the UCS-World (probably is current) and make layer SOLIDS current.
2. Enter paper space with **PS<R>** – or menu bar selection.
3. Use the ZOOM command and zoom in on a window from (5,20) to (195,150).

4. Return to model space – **MS<R>**.
5. SOLMESH the house then SHADE.
6. From the menu bar select **UTILITY**
   SLIDES
   MSLIDE
   AutoCAD prompts with the Create Slide File dialogue box
   respond **pick Type it**
   AutoCAD prompts `Slide File......`
   enter **\SOLID\SL1<R>**
7. Enter REGEN<R> to remove the shading.
8. Using the Layer Control dialogue box, freeze layer ROOF.
9. SHADE the house again.
10. Enter **MSLIDE<R>** at the command line, and from the dialogue box pick Type it and enter the slide name \SOLID\SL2.
11. Now REGEN and SOLWIRE the composite to return to wire-frame representation.

Adding detail

Furniture will now be added to the house in the form of solid box primitives so

1. Still in model space, lower left viewport active, zoomed view?
2. Return to the previous view with **PS<R>**
   ZOOM P<R>
   MS<R>.
3. Make a new layer FURN, colour blue and current.
4. Use SOLBOX with the following entries:
   (a) table primitive   (b) chair primitive
       corner of box: 40,70,10   corner of box: 50,45,10
       other corner: 160,140,10   cube option: C
       height: 50   length: 25

5. The small chair box is to be arrayed about the table, so select **Construct-Array**, then
   (a) pick the small box as the object (lower right viewport easier)
   (b) enter R for a rectangular array
   (c) enter two rows and two columns
   (d) enter a row distance of 95
   (e) enter a column distance of 75
6. Union the four chairs with the table – I found it easier to use the lower right viewport to select the objects.
7. Create a sofa with SOLBOX and the following entries:
   (a) corner of box: 210,150,10   (b) corner of box: 210,40,10
       other corner: 280,165,10   other corner: 280,25,10
       height: 30   height:30
   (c) corner of box: 280,40,10   (d) corner of box: 265,40,10
       other corner: 265,150,10   other corner: 225,150,10
       height: 50   height: 25
8. Union these four primitives.
9. We now have four separate composites in our house:
   (a) the walls of the house
   (b) the roof and chimney – presently frozen
   (c) a table and four chairs
   (d) a sofa.
10. Your drawing should now resemble Fig. 27.1.
11. At this stage save as **\SOLID\HOUSE** (or a new name?)
12. SOLMESH the composite then SHADE in each viewport – quite a nice display?
13. Enter the following sequence
   (a) lower left (3D) viewport active
   (b) enter paper space
   (c) zoom a window as before, i.e. (5,20) to (195,150)
   (d) enter model space
   (e) SHADE again if necessary
   (f) MSLIDE with the name **\SOLID\SL3**.
14. REGEN to remove shading, but stay in the zoomed 3D view.

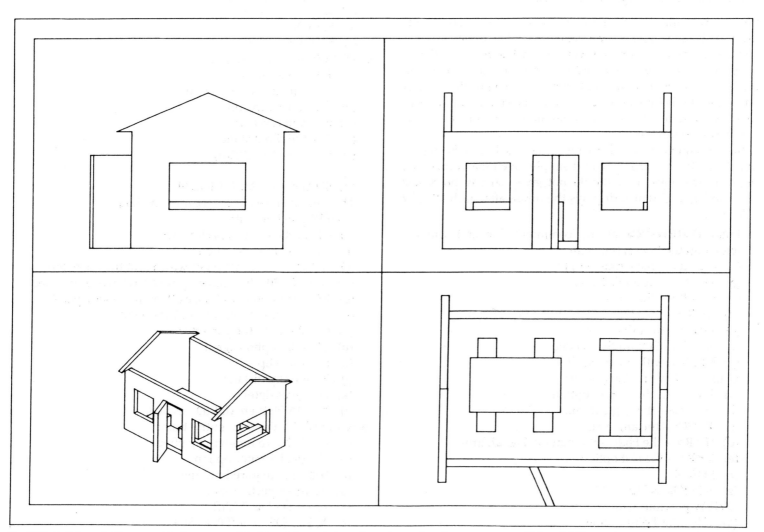

Fig. 27.1. Solid model house with furniture added (roof frozen) (see also colour plates between pages 200 and 201).

Dynamic viewing the house

We will dynamic view the house in the zoomed viewport with the roof layer frozen. It was stated that the dynamic view options can be used consecutively, and while this is perfectly true, we will use the command six times instead of one. The reason for this is that we want to make a slide of the various entries, and it is thus necessary to quit the command in order to make the slide. This involves a bit more work for us, but it is not too much trouble.

In the sequences which follow, I have deliberately missed out the AutoCAD prompts, and simply given the entries as a list of instructions. This should not give you any problems! Refer to Fig. 27.2 which displays the house after each DVIEW selection.

1. Enter **DVIEW<R>** at the command line and window (crossing) the house, then enter:
    (a) **CA<R>** – the CAmera option
    (b) **60<R>** – the angle from......
    (c) **120<R>** – the angle in...
    (d) **H<R>** – the Hide option giving Fig. 27.2(a)
    (e) **X<R>** – the exit option
    (f) SHADE – SOLMESH needed?
    (g) MSLIDE with the name SL4
2. Using DVIEW previous, enter:
    (a) **TA<R>** – the TArget option
    (b) **-40<R>** – the angle from.....
    (c) **40<R>** – the angle in....
    (d) **H<R>** – the Hide option giving Fig. 27.2(b)
    (e) **X<R>** – the exit option
    (f) SHADE
    (g) MSLIDE as SL5.
3. DVIEW previous with:
    (a) TW – the TWist option
    (b) 180 – the new twist angle

(c) H – gives Fig. 27.2(c)
(d) X
(e) SHADE
(f) MSLIDE as SL6.
4. DVIEW previous with:
    (a) TW – twist again
    (b) 0 – restores house correctly
    (c) CL – the CLip option
    (d) F – front clip entry
    (e) 30 – the clip distance
    (f) H – gives Fig. 27.2(d)
    (g) X
    (h) SHADE then MSLIDE as SL7.
    This option allows us to 'see into' the house.
5. DVIEW previous with :_
    (a) CL, F, 200 – removes clip effect
    (b) PO – the POints option
    (c) 160,0,0 – the target point which is at the open door
    (d) @50,–50,50 – the camera position relative to the target
    (e) PA – pan option as house off-centre in viewport?
    (f) 160,0,0 – the displacement base point
    (g) @100,–50,0 – the pan effect
    (h) CL – clip option again
    (i) F – front clip
    (j) 0 – the clip distance
    (k) H – gives fig(e)
    (l) X, SHADE then MSLIDE as SL8.
6. Last DVIEW previous entry, then:
    (a) CL, F, 200 to remove clip effect
    (b) D – the Distance option
    (c) 800 – the required distance
    (d) new perspective icon
    (e) H – gives Fig. 27.2(f)
    (f) X, SHADE, MSLIDE as SL9.

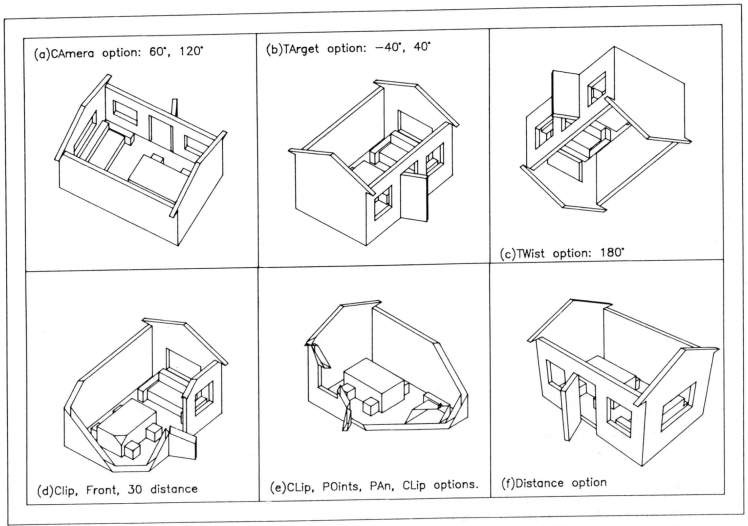

Fig. 27.2. DVIEW options with house composite solid model (see also colour plates between pages 200 and 201).

Notes

1. When DVIEW is used, the command gives a **ghost image** when the required component is selected. This image allows the user to 'see' the component before entering the required values. Values can be entered from the keyboard (as our examples) or simply by picking when the image gives the display required.
2. Several of the commands give a **slider bar** while others give prompts in the co-ordinate area of the menu bar. When using DVIEW I generally watch the display until it is roughly in the orientation I require then enter a value from the keyboard approximately the same as that displayed in the menu bar.
3. Entries with DVIEW are always relative to the WCS and not the UCS.
4. Toggle means moving ghost image then **<R>**.

The DVIEW options

DVIEW is very powerful and gives the user a command with options unavailable with any other command. The result of options such as twist and clip are very useful (try turning a 3D solid model upside down using conventional commands), while the distance option adds real perspective to the component. A brief description of the options will now be given.

Camera: this option is very similar to the VPOINT 'R' command, which zooms a new view to the drawing extents. With the camera option of DVIEW, the user has control over the view display. The view of the component with this option is:

+ from angle – viewed from above
– from angle – viewed from below
+ in angle – viewed from front right side
– in angle – viewed from front left side.

The camera angles have restrictions on the value entered:
(a) angle from value must be between –90° and 90°
(b) angle in value must be between –180° and 180°.

Target: a similar option to camera and:

+ from angle – viewed from underneath
– from angle – viewed from above
+ in angle – viewed from rear left side
– in angle – viewed from rear right side.

The same result can be obtained with the camera and target options of the DVIEW command. Camera angle inputs of 10 and 50 will give the same view as target angle inputs of –10 and –130. The reason for this is simple, as the target option tilts the component in the 'opposite sense' to the camera tilt.

The target angles have the same restrictions as the camera angles.

Distance: this option changes the display from parallel to perspective view and the distance referred to is the distance from the camera to the target. The perspective icon (an oblong box) is obtained as is the slider bar. The bar can be confusing to new users of the DVIEW command, as it does not give actual distances, but 'scaled values' relative to the default, i.e. the value in the < > brackets. These scaled values are graded as follows:

0xdefault – at left side of slider
1xdefault – at first line (the left one)
4xdefault – at second line (middle)
8xdefault – at right line
16xdefault – at right side of slider bar

I prefer to enter a distance value from the keyboard. When the distance option is used, components stay in their VPOINT orientation. If the DVIEW command is ended with the distance option, pan and zoom are not then permitted in a perspective viewport, although they are options of the DVIEW command itself.

Points: allows the user to specify a point on the target and the position of the camera. Absolute or relative co-ordinate entry is permissable, but can be confusing due to the orientation of the component.

Clip: a very useful option as it allows the user to remove objects from the front and back of the component being viewed by the insertion of a 'clipping plane'. This may be necessary if a 3D component has other objects on the same layer and this layer cannot be frozen. The effect of clip is such that:

(a) a back clipping plane 'hides' all objects that are behind it
(b) a front clipping plane 'hides' all objects in front of it.

The actual clipping plane is relative to the camera position, and the clip plane is perpendicular to the line of sight between the camera and the target. Clipping is possible in both parallel and perspective projection. A common use for the clip option is to 'cut-away' walls to 'see inside' a component – as our example. This may be important with solid modelling as the user may want to see detail which cannot be obtained by other means.

Twist: this option rotates the view about the line of sight by the specified angle, and allows components to be 'turned upside down' fairly easily.

Pan: similar to the pan command which cannot be used in a perspective viewport.

Zoom: the ZOOM command cannot be used in a perspective viewport and the zoom option allows the user to 'change the field of view' by changing the 'zoom lens' of the camera. The image can be scaled without altering the camera and target points. The default zoom lens length is 50 mm (scale factor 1) and:

(a) zoom scale >1 – increases image size
(b) zoom scale <1 – decreases image size.

The actual effect of altering the zoom factor is:

(a) increased lens length gives a telephoto effect
(b) decreased lens length gives a wide-angle effect.

Hide: gives hidden line removal while still in the DVIEW command.

Undo: undoes the last option and keeps the user in the DVIEW command. This option can be used until all options entered are undone.

Exit: quits the DVIEW command, the component being displayed with the effect of the last option used.

Creating a slide show

In the investigation of the DVIEW command, we made nine slides (SL1–SL9). These slides can be individually viewed by entering VSLIDE at the command line and selecting the required slide name from the slide file dialogue box. Slides can also be combined and run as a 'slide show'. This requires a script file to be written by the user. Script files are text files and can be written using any word-processor, but can also be written while in AutoCAD and this is the method we will adopt.

1. From the menu bar select **UTILITY**
   External or **Commands**
   SHELL

AutoCAD prompts	with the text screen
and	OS Command_
enter	**edlin C:\SOLID\SHOW.SCR<R>**
AutoCAD prompts	New file
then	*
enter	**1i<R>**
AutoCAD prompts	1:*_ and enter VSLIDE SL1**<R>**
AutoCAD prompts	2:*_ and enter DELAY 500**<R>**
AutoCAD prompts	3:*_ and enter VSLIDE SL2**<R>**
AutoCAD prompts	4:*_ and enter DELAY 500**<R>**
AutoCAD prompts	5:*_ and enter VSLIDE SL3**<R>**
AutoCAD prompts	6:*_ and enter DELAY 500**<R>**
AutoCAD prompts	7:*_ and enter VSLIDE SL4**<R>**
AutoCAD prompts	8:*_ and enter DELAY 500**<R>**
AutoCAD prompts	9:*_ and enter VSLIDE SL5**<R>**

AutoCAD prompts	10: *_ and enter DELAY 500<R>
AutoCAD prompts	11: *_ and enter VSLIDE SL6<R>
AutoCAD prompts	12: *_ and enter DELAY 500<R>
AutoCAD prompts	13: *_ and enter VSLIDE SL7<R>
AutoCAD prompts	14: *_ and enter DELAY 500<R>
AutoCAD prompts	15: *_ and enter VSLIDE SL8<R>
AutoCAD prompts	16: *_ and enter DELAY 500<R>
AutoCAD prompts	17: *_ and enter VSLIDE SL9<R>
AutoCAD prompts	18: *_ and enter DELAY 500<R>
AutoCAD prompts	19: *_ and enter RSCRIPT<R>
AutoCAD prompts	20: *_ and CTRL C
AutoCAD prompts	*_
enter	e<R> to save the text file and return to AutoCAD.

2. From the screen menu select **UTILITY**
   SCRIPT

| AutoCAD prompts | with the Select Slide File dialogue box |
| respond | **pick SHOW then OK** |

3. All being well a slide show of your house with the DVIEW options will be shown on the screen. This can be stopped at any tie with CTRL C.

Note

1. Slides are made with MSLIDE and viewed with VSLIDE.
2. Slides can be included in a script file to give a slide show. A script file is a text file with the extension **.SCR**.
3. Delays have been added to the slide show to 'slow down' the slides being shown. The greater the number with DELAY, then the longer the delay.
4. RSCRIPT in the last line re-runs the slide show.
5. REDRAW will restore the drawing screen.

❏ *Summary*

1. DVIEW is a very powerful (and under-used) AutoCAD command.
2. DVIEW is very suited to solid modelling when ordinary commands may not give the expected result.
3. The clip option is very useful.
4. The distance option will result in the component being displayed in perspective view.
5. DVIEW requires practice and patience, and I would recommend that users who are not familiar with the command spent some time with it until they are competent at using the options.
6. The slides were an afterthought!

An extra solid model composite for DVIEW

The house example which was used to investigate the DVIEW command was a reasonable model, but the effect of the clip option was rather obvious, i.e. removing walls to 'see into' the room. To reinforce the advantage of using DVIEW with solid models, we will create a further composite consisting of several primitives. The entries will be given as a sequence of user inputs, and you should reason out all the entries.

Creating the composite

1. Open your SOLA3 standard sheet in model space with layer SOLIDS current and the lower left viewport active.
2. SOLBOX with corner at 0,0,0; (C)ube option; side length 300. The colour is to be red – should be?
3. Centre each viewport with ZOOM C at 150,150,150 and 400 magnification, 600 magnification in the 3D viewport.
4. UCS-Z at 45° to rotate the UCS about the Z axis.

5. SOLWEDGE with corner at 0,0,0; length 60; width 200; height 100 and colour yellow.
6. Move the yellow wedge from 0,0 by @0,–100.
7. Polar array the yellow wedge about the point 213.5,0 for four items with 360 fill and rotation.
8. Subtract the four yellow wedges from the red cube.
9. Restore UCS-World.
10. Create a box and sphere inside the red cube with:

SOLBOX | SOLSPHERE
corner: 50,50,250 | centre: 150,150,150
length: 200 | radius: 110
width: 200 | colour: blue
height: –150 |
colour: green |

11. Subtract the blue sphere and green box from the red cuboid which gives an unusual effect?
12. SOLCYL with centre at 150,150,0; radius 20; height 310 and colour yellow.
13. Subtract the yellow cylinder from the composite.
14. SOLCYL with centre at 0,0,0; radius 20; (C)entre option with @150,150,150 as the centre of the other end. This cylinder is to be yellow.
15. **Array 3D** this yellow cylinder (Last option); Polar; four items; 360 fill; with 150,150,150 as the first point and 150,150,300 as the other point of array.
16. Subtract the four yellow cylinders from the composite (lower right viewport easier for selection).
17. The composite is now complete, and consists of an external cube component with some interesting bits taken from its inside.
18. What do you think shading will give?
19. SOLMESH then SHADE the composite in each viewport, which gives the red cube shape with a few yellow parts – wedge and holes.

This is as expected(?) as all the work is 'inside' the cube and hence difficult to 'see'.

20. REGENALL and SOLWIRE in preparation for DVIEW.
21. Figure 27.3 shows the composite with HIDE, and as expected it does not show what is inside the cube.
22. Save your model as **SOLID\?????**.

DVIEWing the composite

1. In each viewport enter the sequence:
   (a) DVIEW
   (b) pick the composite
   (c) CL – clip option
   (d) F – front option
   (e) 275 – the clip distance.
   (f) X – to exit command.
2. SOLMESH and SHADE each viewport.
3. Figure 27.4 gives the effect of this clip option, which is plotted with HIDE.
4. Some of the viewports do not show any effect of the DVIEW option entered. Any reason for this?
5. Open your saved drawing of the composite **SOLID\????**.
6. In each viewport set the viewpoint to VPOINT 'R' 315, 30 and centre each viewport with ZOOM C at 150,150,150 with 500 mag.
7. Using DVIEW-CL-F, enter the following clip distances:
   (a) lower left: 300
   (b) upper left: 250
   (c) upper right: 200
   (d) upper left: 150.
8. The result of these entries is interesting as you can now 'see' what is happening 'inside the cube'
9. Figure 27.5 displays these entries with HIDE on.
10. This NOW completes the DVIEW chapter on solid models.

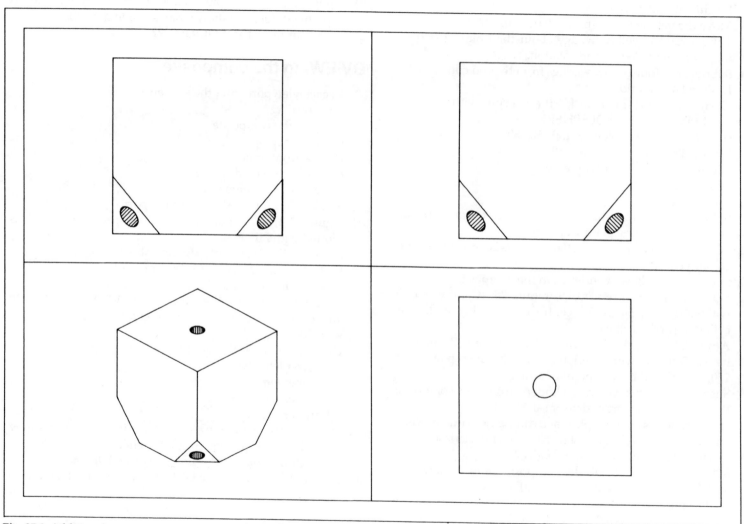

Fig. 27.3. Additional component to investigate the DVIEW command.

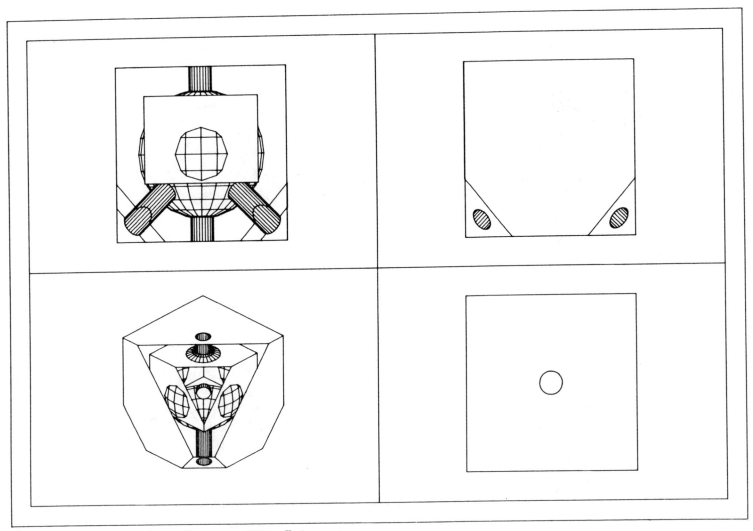

Fig. 27.4. DVIEW command with CLip, Front, 275 in all viewports.

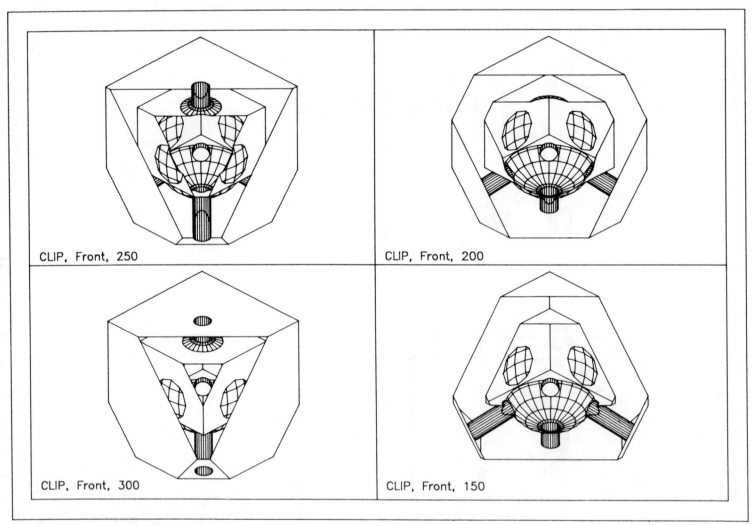

Fig. 27.5. Different DVIEW CLip, Front with the same VPOINT (see also colour plates between pages 200 and 201).

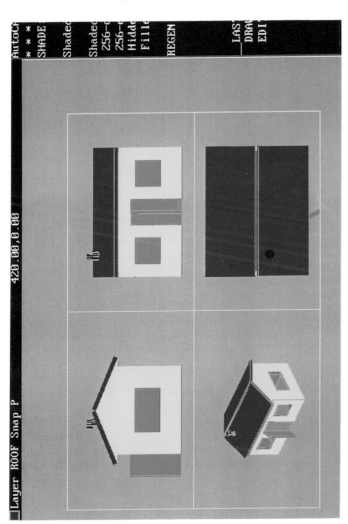

Fig. 26.2. Solid model house with roof, chimney and open door.

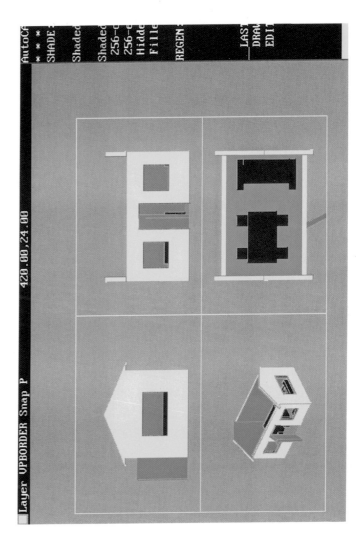

Fig. 27.1 Solid model house with furniture added (roof frozen).

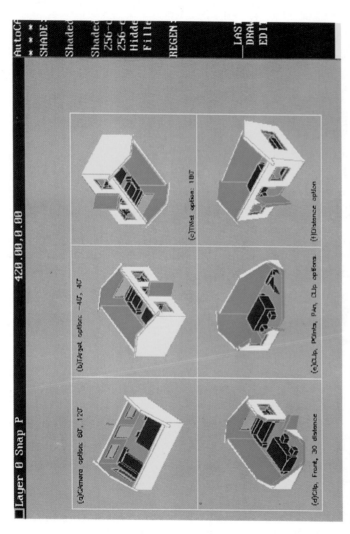

Fig. 27.2. DVIEW options with house composite solid model.

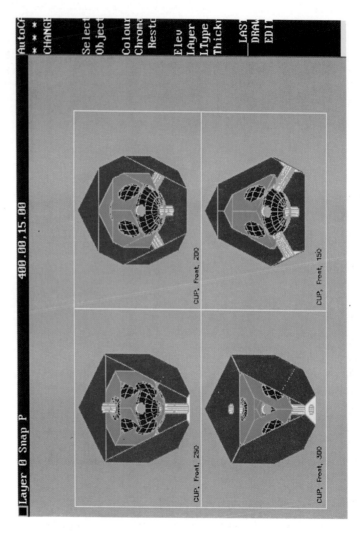

Fig. 27.5. Different DVIEW CLip, Front with the same VPOINT.

28. Rendering solid models

Rendering with AutoCAD will allow 3D drawings to be converted into coloured images, the user having control over the illumination which is added to the display. As with shading, the only objects which can be rendered are

- extruded components
- 3D faced and meshed objects
- solid models.

Rendering will turn a coloured shaded drawing into a display which will impress any employer and/or customer and is relatively easy to apply. It adds that 'little bit extra' to the final drawing.

This chapter will investigate the RENDER command inherent within AutoCAD and will not consider:

(a) AutoSHADE: an add-on which greatly improves the render effect
(b) 3D Studio: a separate software package which allows the user to render, animate and produce films (well nearly).

Render has its own terminology, for example lights, scenes, finishes, and so on, and we will investigate each during this chapter. The options available with Render can be obtained from:

(a) the screen menu by picking **RENDER**
(b) the menu bar with **Render**

Both selections give the same options, although the order of display is different:

Screen RENDER	*Pull down Render*
RENDER	Render
HIDE	Shade
DDVIEW . . .	Hide
LIGHT . . .	————
SCENE . . .	Views . . .
FINISH . . .	————
RPREP . . .	Lights . . .
STATS . . .	Scenes . . .
REPLAY . . .	Finishes . . .
SAVEIMG.	————
	Preferences . . .
Unload	————
————	Statistics
RMAN	Files >
	————
	Unload Render
	Renderman . . .
	————

Remember that menu items with (. . .) after their name result in dialogue boxes, and it is obvious that Render has several new dialogue boxes. These will be discussed (and displayed where appropriate) as we work through the exercises.

The following may assist the user with some of the terminology:

Render: (root menu or menu bar) displays the Render menu.

RENDER: selected from the Render menu (or entered from the keyboard) will perform the render function. Render is an AutoCAD ADS application.

Renderman: an additional package supplied with AutoSHADE and is not considered in this chapter.

AVERENDER: the AutoCAD Render program file.

Loading AutoCAD render

As Render is an ADS application it is not instantly available to the user, but must be loaded before it can be used – similar to AME. The 'render package' is automatically loaded the first time any of the render options are used, but it can also be pre-loaded ready for use at some later time. The sequence to load ADS Render prior to its use is to load AutoCAD. Select from the menu bar **File**

> **Applications...**

AutoCAD prompts	with the Load AutoLISP and ADS File dialogue box
respond	**pick File...**
AutoCAD prompts	with the Select LISP/ADS Routine dialogue box
respond	**pick AVERENDR.EXP** from the files list then **OK**
AutoCAD prompts	with the Select LISP/ADS Routine dialogue box
and	C:\ACAD12\AVERENDR.EXP added and highlighted in blue

(note that ACAD12 is my directory name for AutoCAD)

respond	**pick Load**
AutoCAD prompts	with the text screen
and	AVE_RENDER is not yet configured
then	Select rendering display device

1. AutoCAD's configured P386 ADI combined display.....
2. P386 AutoDESK Device Interface....
3. None(Null rendering driver)
Rendering selection<1?

enter	**<RETURN>**

AutoCAD prompts `Default <1> selected`
Do you want to do a detailed configuration...

respond	**<RETURN>**, i.e. default **<N>** selected.

AutoCAD prompts `Select mode to run display/ rendering combined driver`

1. Render to display viewport.
2. Render to rendering screen.
AutoCAD driver does not support rendering to a viewport
Rendering screen automatically selected
Press RETURN to continue

respond	**<RETURN>**

AutoCAD prompts `Select rendering hard copy drive`

1. None (Null render device)
2. P386 AutoDESK Device Interface rendering driver
3. Rendering file (256 color map)
4. Rendering file (continue color)
Rendering hard copy selection<1>

respond	**<RETURN>**

AutoCAD prompts `Default <1> selected`
Enable handle...done
Initialising preferences...done
File C:\ACAD12\AVERENDER.EXP loaded
then Returns the command line prompt

respond	**F1** to flip back to drawing screen.

Once the above sequence has been completed, render can be used. If you decide not to follow the sequence but simply to load the 'render package' when you select a render option, you will still have to 'plod through' the list of selections, so you will not save much time whatever method you decide to use. My preference is to pre-load, then proceed with the drawing.

Note

1. I would imagine that most users would simply accept the default values to the prompts. Your system may support some of the other devices listed, but for our investigation, the defaults are suitable.
2. AutoCAD render can be 'unloaded' (if you want to) by one of three methods:
   (a) selecting Unload... from the RENDER screen
   (b) selecting Unload from the Application dialogue box
   (c) entering (Xunload 'averender') at the command line.

AutoCAD's render 'pictures'

We will now render two drawings supplied with AutoCAD and which should (hopefully) be in your system.

1. Load AutoCAD (if needed) then select **File**
   > **Open**
   > **TUTORIAL** directory
   > **KITCHEN2** drawing

AutoCAD prompts	with a multi-screen kitchen layout
respond	ensure large viewport is active
then	select from the menu bar
	Render
	Render
AutoCAD prompts	by loading the render package (takes a few seconds)
then	five **<RETURNS>** to prompts needed
then	a coloured display of the kitchen is displayed.

This coloured image is in perspective.

respond	**<RETURN>** to return to drawing screen.

2. Select **File**
   > **Open** – discard changes
   > **PINS2** drawing – TUTORIAL current directory?

AutoCAD displays a multi-screen drawing. **RENDER** the large viewport, then **<RETURN>** to the drawing screen.

3. The two coloured images obtained with render are impressive, and after the worked examples are investigated, you will have the ability to produce similar images.

Render example

Our example for render will be the solid model house which was created for the shading and dynamic viewing chapters, so

1. Open your **HOUSE** drawing with the furniture added.
2. Lower left viewport active in model space with layer SOLIDS current. Thaw layer ROOF and set UCS-World, i.e. origin is at the front lower left corner of the house.
3. Pre-load render if needed.
4. SOLMESH the complete house which will load AME.
5. Select from the menu bar **Render-Render** and AutoCAD prompts:

```
Using current view
Default scene selected
Projecting objects into view plane
1 solid selected
Surface meshing of current solid is complete
...............
.........................................
Applying parallel projection
Calculate extents for faces
Sorting 408(?) triangles by depth
Checking 408 triangles for obscuration
Calculate shading and assign colors
Outputting triangles.
```

6. Your solid model house will be displayed as a coloured image with a black background. The effect is quite impressive, as we have not yet added lights to the model.
7. Keyboard **<RETURN>** to return to the drawing screen.

Note

1. Any viewport can be rendered, although it is usual to render a 3D viewport. Try another viewport render if you want.
2. Enter paper space with **PS<R>** and then enter **RENDER<R>**
   prompt   **Command not allowed in Paper Space**, i.e. render is a model space command.
3. Return to model space, and we will now investigate how lights can be added to our model.

Lights

To obtain the full benefit of rendering a drawing, AutoCAD allows the user to add lights. There are two types of light available:

(a) distant light: is light giving parallel rays in a direction specified by the user, e.g. sunlight shining on an object is a distant light
(b) point light: is light which comes from a point source and 'shines' in all directions, e.g. a light bulb can be considered as a point light.

Note in the light dialogue box a third type of light source is mentioned – spotlight. This type of light is not available to 'ordinary' AutoCAD rendering, but is available with full AutoSHADE (Renderman). It will not be considered in this chapter.

Lights are positioned in a drawing by the user. Distant lights require a target position and a location position, similar to the POints option of the DVIEW command, while the point light requires a light location. It is normal to use co-ordinate entry for the light positions. The user can also control the light intensity (point lights) and the ambient factor (distant lights). Lights are given names, allowing them to be repositioned and 'turned off' as required. Point and distant lights are represented by symbols (icons), these icons being added to the drawing at the co-ordinate entry specified by the user. The symbols are shown in Fig. 28.1.

1. Return to your house drawing – lower left viewport.
2. From the menu bar select **Render**
                         Lights...

prompt	with the Lights dialogue box – Fig. 28.2
respond	(a) select Inverse Linear
	(b) **pick New...**
prompt	with the New Light Type dialogue box – Fig. 28.3
respond	**pick Distant Light then OK**
prompt	with the New Distant Light dialogue box
respond	(a) enter **D1** as Light name – Fig. 28.4
	(b) **pick Modify**
prompt	Enter light target<current>
enter	**0,0,0<R>**, i.e. origin position
prompt	Enter light location
enter	**@0,–50,10<R>** – relative to origin
prompt	New Distant Light dialogue box
respond	**pick Show...**
prompt	with Show Light Position dialogue box – Fig. 28.5
respond	**pick OK**
prompt	New Distant Light dialogue box and pick OK
prompt	Lights dialogue box and pick OK.

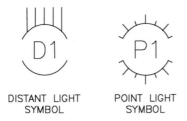

DISTANT LIGHT
SYMBOL

POINT LIGHT
SYMBOL

Fig. 28.1. Light symbols.

3. Now set the following new lights:
   A. (a) Distant light type
      (b) Light name: D2
      (c) Modify with target position: 150,100,200 location position: @0,0,300
   B. (a) Point light type
      (b) Light name: P1
      (c) Intensity: about 300 – use slider bar
4. We now have three lights:
   (a) D1: distant light at point 0,–50,10 shining on point 0,0,0
   (b) D2: distant light at point 0,0,500 shining on point 150,100,200
   (c) P1: point light positioned at point 150,100,200 (on 'ceiling')
5. Save your drawing (with lights) as ?????.

Notes

1. Lights can be point or distant.
2. Lights are positioned by the user.
3. The point light intensity can be altered as required.
4. The ambient setting for distant lights can be altered by the user.
5. Light positions can be altered by the user using Modify.

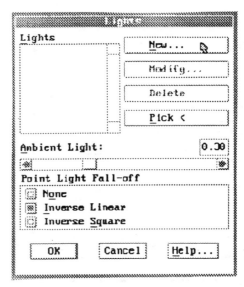

Fig. 28.2. Lights dialogue box.

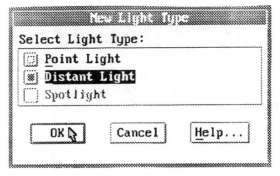

Fig. 28.3. New Light Type dialogue box.

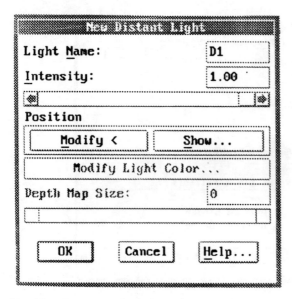

Fig. 28.4. New Distant Light dialogue box.

Show Light Position

Location	Target
X = 0.00	X = 0.00
Y = -50.00	Y = 0.00
Z = 10.00	Z = 0.00

OK   Help...

Fig. 28.5. Show Light Position dialogue box.

Scenes

Lights which have been added to a drawing are used with scenes to give the desired render effect and *a scene is a view with lights added.*

The above statement requires the user to set a view, i.e. the VIEW command must be used, so

1. Ensure lower left viewport active.
2. From the screen menu select **DISPLAY**
   VIEW:

Prompt	?/Delete/Restore/Save/Window
enter	**S<R>**
Prompt	View name to save
enter	**VIEW1<R>**
Prompt	with command line.

3. From menu bar select **Render**
   Scenes...

Prompt	with the Scenes dialogue box
respond	**pick New...**
Prompt	with New Scene dialogue box
respond	(a) enter SCENE1 at Scene name box
	(b) pick VIEW1
	(c) pick light D1 – Fig. 28.6
	(d) pick OK
Prompt	with the Scenes dialogue box
respond	using the New... option, create scenes using the following information:

Scene name	View used	Lights selected
SCENE2	VIEW1	D2
SCENE3	VIEW1	P1
SCENE4	VIEW1	D1 and P1
SCENE5	VIEW1	D2 and P1
SCENE6	VIEW1	*ALL*

4. To 'see' a rendered scene, enter **SCENE<R>** at the command line:

Prompt with Scenes dialogue box
respond **pick SCENE1 then OK**
Prompt with command line
enter **RENDER<R>**

5. Your house will be displayed as a coloured image for the light D1.
6. Using the SCENE/RENDER sequence, select the other scenes to see the render effect of the various added lights. The effects are quite interesting.
7. Save your drawing with the created scenes.
8. This ends our brief investigation into rendering solid models.

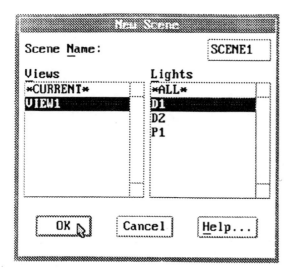

Fig. 28.6. New scene dialogue box.

Notes

1. Scenes are created from views.
2. Scenes are views with lights added.
3. Different views can be used for scenes.
4. A view can be used for several different scenes.
5. The user controls the lights which are added to a scene.
6. SCENE is an AutoCAD command used prior to RENDER.
7. The lights added to a scene can be altered using Modify from the Scenes dialogue box. This allows the user to change:

   (a) the view selection
   (b) the lights in the scene

Images

Rendering a screen or a scene produces a coloured image of the viewport and this image can be saved as a file. Saved images can replayed at any time and imported into desk-top publishing software. The procedure is relatively simple, the user:

(a) selecting **Render-Files>-Save Image...**
(b) selecting the file format from the Save Image dialogue box – usual format is TGA
(c) ensuring directory name is correct
(d) entering the image name, e.g. HOUSE_1 or similar
(e) picking OK.

The sequence will then save the active viewport as a TGA file. Saved images can be replayed by selecting **Replay...** and selecting the required image name from the resultant dialogue box.

Saving images is beneficial as they can be replayed at any time and the original drawing they were created from does not need to be current, i.e. the user can build up a library of images for replay at a later time.

Finishes

The finish option allows colours and finishes to be assigned to objects in a drawing. This gives shiny and dull appearances. Using finishes requires some 'trial and error' until the correct setting is obtained. The New Finish dialogue box allows the user four settings, these being:

(a) ambient (0-1)   controls how light is reflected off surfaces and:
0 – no ambient light reflected
1 – all ambient light reflected

(b) diffuse (0-1)   determines how much light is reflected from surfaces and: 0 – shiny; 1 – dull

(c) specular (0-1)   controls the shiny surface reflection versus the diffuse reflection

(d) roughness (0-1)   controls how widespread the specular setting is.
High roughness values make the specular highlight larger.

Note

1. Generally: **AMBIENT+DIFFUSE+SPECULAR=1**.
2. Using the finish option requires some work from the user.

29. . . . and finally

The object of this book was to introduce the reader to the concept of solid modelling. If you have worked through all the exercises, you should have a sound grasp of the principles involved in creating solid composites. The chapters on shading, dynamic viewing and rendering were added as an afterthought and should make you aware of the power of creating drawings as solid models. Most drawings can be displayed in multi-view format as solid models, although it may take some time to work out the primitives which are required.

Solid modelling becomes easier with practice. If you are producing CAD drawings, try and make the component as a solid model either as a series of solid primitives or by using the swept primitives which are generally underused.

I hope that you have enjoyed working through the book and any comments you have on how to improve the exercises would be more than welcome.

Good luck!!

Bob McFarlane

Robert McFarlane

Index

2D models   1
3D models   1

ACAD.MAT   118, 119
Ambient   208
Ambiguity   1, 3
Attach   161
Applications   202
ARRAY   25, 60, 73, 190
Array 3D   101, 157
Auxiliary view   146, 153
AVERENDER   202

Balanced tree   55, 56, 57
Baseplane   31
BLOCKS   161
Boundary representation   2

CAD model   1
CAMERA   189, 194
Chamfered edges   43
Change primitive   101
CHPROP   14, 21, 23, 25, 28, 73,
    80, 96, 148, 153, 182
Classic geometry   83
CLIP   189, 195
Composite   14, 15, 16, 53, 55, 99,
    106, 122, 134, 146
Composite regions   84
Computer models   1

Construct   101, 157, 190
Constructive Solid Geometry   2
COPY   76, 78, 92, 106
CSG   2, 55, 84
CSG units   117
Cutting solids   55

DDSOLMASP   117, 120, 129
DDSOLMAT   81, 129
DELAY   195
Density   129
Dialogue boxes
    create mass property file
        121
    hatch parameters   115
    lights   205
    mass properties   118
    material browser   121, 129
    new distant light type   206
    new light type   205
    new scene   207
    other parameters   115
    select material   121
    show light position   206
    system variables   115
    units   115

Edge   83
Edlin   195
EXPLODE   19, 193

FILLET   105
Filleted edges   43
Finishes   208
Flat models   1

HANDLES   4, 5, 18
Hidden profiles   7
HIDE   15, 23, 25, 28, 32, 36, 43,
    61, 73, 78, 80

Images   207
Interference   1, 2, 161
Intersection   53, 54
INSERT   161
Irregular surfaces   83

LAYER   18, 19
Layer control   4
Lights   204, 205, 206
LIMITS   4, 23
Linear expansion   129
LIST   4, 149
Load   119, 202
Loops   84

Materials   119
Material properties   81
MCS icon   92, 94, 95, 96, 106, 112
MIRROR   32, 41, 69
Model   1

Model space   9, 69, 134
Motion codes   92, 94, 95, 96
MOVE   23, 76, 78, 94, 99, 107,
    124, 149, 153
MPR file   120
MSLIDE   190, 192, 196
MVIEW   84, 92, 149, 162

Named UCS   4, 183

PAN   189, 195
Paper space   9, 69, 134, 189
PEDIT   32, 36, 69
Physical models   1
PLINE   32, 36, 69, 73, 76, 78, 80,
    105, 134, 146, 148
POINT   189, 194
Primitives   14, 16, 21, 53, 55, 60,
    99, 146
PRISMS   12, 21

Ray projection   117, 118
REDRAW   196
REGEN   15, 23, 25, 28, 32, 36, 61,
    73, 78, 80, 136
Region   84, 99, 105
Render   1, 2
RENDER   201, 204, 206, 207
REPORT   117, 120, 124, 125, 126,
    127, 128, 130

ROTATE   23, 94, 186
RSCRIPT   196

Scenes   206
SCRIPT   196
Script file   195, 196
Separation   55
Set view   189
SETUP   116, 122
SHADE   15, 23, 25, 28, 32, 36, 43, 44, 60, 73, 78, 80, 96, 136, 148, 182, 184, 186, 188
SHADEDGE   186, 188
SHADEDIF   186, 188
SHELL   195
SLIDES   190
Slide show   195
Solid model   2, 3, 16, 55
Solid modeller   2
Solid model commands
   SOLAREA   73, 78, 112, 116, 117, 120, 122
   SOLAXCOL   112, 114
   SOLBOX   15, 21, 60, 76, 99, 101, 136, 182, 183, 187
   SOLCHAM   43, 51
   SOLCHP   99, 101, 103, 105, 106, 131, 137, 162
   SOLCONE   25
   SOLCUT   55
   SOLCYL   14, 15, 25, 60, 73, 80, 99, 105, 186, 197
   SOLDECOMP   113, 114, 122
   SOLDELENT   114
   SOLDISPLAY   112, 114
   SOLEXT   32, 41, 78, 80, 84, 148
   SOLFEAT   137, 144, 153
   SOLFILL   44, 51

SOLHANGLE   84, 105, 113, 114, 131, 139, 166
SOLHPAT   84, 105, 113, 114, 131, 139, 166
SOLHSIZE   84, 105, 113, 114, 131, 139, 166
SOLINT   53, 60
SOLLENGTH   112, 114
SOLLIST   12, 14, 15, 23, 43, 44, 60, 80, 81, 116, 120, 122, 129, 161, 163
SOLMASSP   73, 78, 80, 81, 113, 116, 117, 120, 122
SOLMAT   116, 118, 120, 124
SOLMATCURR   113, 114
SOLMESH   15, 23, 25, 28, 32, 36, 43, 44, 61, 73, 78, 80, 96, 136, 137, 138, 148, 184, 197
SOLMOVE   92, 94, 96, 98, 105
SOLPROF   18, 19, 139, 144, 153
SOLRENDER   112
SOLREV   36, 41, 78, 80, 84, 148
SOLSECT   131, 139, 144, 166
SOLSECTYPE   113, 114
SOLSEP   55
SOLSOLIDIFY   112, 114
SOLSPHERE   28
SOLSUB   12, 14, 15, 33, 85, 122, 182, 183
SOLSUBDIV   113
SOLTORUS   28
SOLUNION   53, 183, 184
SOLVAR   112, 113
SOLVOLUME   112, 114
SOLWDENS   8, 15, 69, 73, 105, 112, 114, 157
SOLWEDGE   15, 23, 76, 148, 183, 197

SOLWIRE   15, 25, 28, 32, 36, 43, 44, 61, 73, 78, 80, 96, 136, 148
Solid model variables   106
Solidify   85
Source solid   53
Specular   208
STRETCH   69, 76, 149
Subtraction   53, 54, 122
Surface model   2, 3
Swept primitive   69, 83

TARGET   189, 194
Thin shells   83
Tilemode   4, 11
TRIM   36
TWIST   189, 195
TYPE   120

UCS   4, 11, 19, 28, 36, 60, 61, 69, 73, 78, 80, 131, 134, 136, 137, 139, 144, 146, 148, 183, 184
UCSICON   60, 136, 146, 166, 182
Unbalanced tree   55, 56, 57
Union   53, 54, 80, 122, 157
UTILITY   190, 195, 196

VIEW   137, 206
Viewport specific layers   7
Visible profile   7, 18
VPLAYER   7, 8, 11, 149
VPOINT   4, 5, 84, 85, 192
VSLIDE   195, 196

Wedge   14, 136
Wire-frame model   1, 3
Write to file   120

Xref log file   166
X refs   157, 161, 162, 166

Young's modulus   129

ZOOM   189
ZOOM C   12, 21, 23, 25, 28, 32, 35, 41, 43, 44, 60, 76, 80, 99, 101, 105, 134, 139, 146, 157, 162, 197
ZOOM XP   11, 28, 32, 35, 149, 153